Learning Mathematics through Modelling and Simulation

An Investigative Approach

Jonaki Ghosh
Associate Professor, Lady Shri Ram College for Women, New Delhi

Amber Habib
Professor, Shiv Nadar Institute of Eminence, Delhi NCR

Geetha Venkataraman
Professor, Dr. B. R. Ambedkar University Delhi

Universities Press

Learning Mathematics through Modelling and Simulation: An Investigative Approach

Universities Press (India) Private Limited

Registered Office
3-6-747/1/A & 3-6-754/1, Himayatnagar, Hyderabad 500 029, Telangana, India
info@universitiespress.com; www.universitiespress.com

Distributed by
Orient Blackswan Private Limited

Registered Office
3-6-752 Himayatnagar, Hyderabad 500 029, Telangana, India

Other Offices
Bengaluru, Chennai, Guwahati, Hyderabad, Kolkata,
Mumbai, New Delhi, Noida, Patna, Visakhapatnam

First published 2023

501876

ISBN: 978-93-93330-38-3

Typeset in URWPalladioL 11pt by 13pt
Textsoft Solutions Private Limted, Hyderabad

Printed at
Yash Printographics, Noida.

Published by
Universities Press (India) Private Limited
3-6-747/1/A & 3-6-754/1, Himayatnagar, Hyderabad 500 029, Telangana, India

Dedicated
to **Peter M Neumann** (1940–2020)
for *chai pe charcha ganit ke saath*
&
to **Sudhansu S Bhattacharjee**
for always being the wind beneath my wings

Preface

Years of teaching mathematics at the school and university levels have made us aware of the lacunae that exist. One of the primary drawbacks of mathematics education, both at the school and undergraduate level, is the fact that mathematics is taught largely as an abstract subject without context, and isolated from its applications in real life. Indeed, at the school level, much of the focus of the curriculum is on problem solving of a routine nature and on applying procedures and algorithms. Prescribed textbooks are often unimaginative and cater primarily to preparation for the examinations. At the undergraduate level, there are many books available that try to bridge the need for abstraction with application. However, when it comes to the pedagogy, needs driven by examinations and completion of syllabi, shift the focus to the theoretical aspects. The applications are treated superficially and appear as window-dressing. Thus, for a vast majority of students, mathematics appears to be dry and uninteresting and this poses a serious impediment to effective mathematics learning.

In response, in the last few decades, many countries across the world have witnessed a major shift of paradigm as far as mathematics education is concerned. Serious experimentation and research related to the integration of technology in mathematics education has indicated the possibility of revolutionising the way students experience the learning of mathematics today. Several studies indicate that appropriate use of technology can transform the mathematics classroom into a dynamic interactive learning environment where the learner can develop a deeper understanding of the subject and the ability to think mathematically, rather than focus on only practising manipulative skills.

In our book, we use and teach three powerful tools that enable a student to independently explore, visualise, conjecture, reason and solve problems. We use the spreadsheet program `Microsoft Excel` to work with situations where the calculations are not complicated in principle, but may be too numerous to carry out easily by hand. `Excel` is extremely easy to pick up. One can think of it as a sheet of paper that interfaces with a scientific calculator. There are also free spreadsheet programs like `LibreOffice Calc` which can be used instead of `Excel` and with essentially the same steps. For visualising and manipulating geometric scenes we use `GeoGebra`, an open source 'dynamic geometry software' (DGS). `GeoGebra` is menu-driven and there is no barrier of having to first learn a particular language to operate it. It can be downloaded to your computer or phone, or opened directly in a browser. Finally, `Mathematica` is a comprehensive 'computer algebra system' (CAS). Its power comes at some cost, of learning how to communicate with it, but the learning curve is not steep and there is excellent built-in help and examples. We use `Mathematica` in almost every chapter of the book. It can be accessed online at `www.wolframcloud.com`, and the free version has minor restrictions which do not matter at our level of use. In Appendix A, we provide a brief introduction to each of these tools. Moreover, whenever we use them, we provide detailed instructions in the text itself.

The purpose of the book is three-fold. Firstly, it may serve as supplementary reading for undergraduate students studying mathematics and mathematical modelling. It aims to familiarise the student with some interesting facets and applications of Calculus, Probability, Number Theory, Linear Algebra and Discrete Mathematics. Secondly, it is a rich source of ideas for investigatory tasks for the enthusiastic senior secondary school student who wishes to explore applications of the mathematical concepts learnt in school. Thirdly, it aims to highlight the role of mathematical software in enabling explorations leading to a deeper insight into the problems. We do not just take recourse to technology as an aid to illustrating the given theory. Instead, technology enables us to generate phenomena and acquire data from which questions naturally arise and create a demand for appropriate theory.

Each chapter focuses on a particular aspect or application of mathematics and comprises a set of investigations, which lead the student to explore the ideas in a sequential manner. The reader will be able to explore the concepts using simple programs and simulations, which are provided as and when necessary. Each chapter includes a set of exercises, called 'Investigations' that prompt analytical thinking, encourage exploration of a concept, or test the students' ability to apply the concept to a realistic problem situation. They ensure that the student learns the necessary procedural skills, which are essential for solving problems in that topic. A section at the end of each chapter provides further references and suggests possible extensions to the topic or problem. Solutions and useful hints for the investigations are given in Appendix B.

Let us briefly survey the contents of the eight chapters. The first two are concerned with gaining insights into Probability through computer simulations. Chapter 1 introduces us to computer-generated 'random numbers' and their use in mimicking chance behaviour. Chance has patterns and these patterns can be put to work. In this chapter, we extract estimates of π from observations about certain random behaviour. We are even able to judge the accuracy of our estimates. In Chapter 2, we have another application of simulation. We use it to explore how some contests or games would play out over many instances.

Chapter 3 introduces the theory of graphs. Graphs are a handy way of organising information so that its patterns become visible, and enable further applications of techniques of counting and algebra. The first portion of this chapter discusses the Königsberg Bridge Problem, the very problem that led to the genesis of graph theory. On the surface this is a geometric problem, but unusual in that geometric notions like length and curvature are irrelevant. Casting the problem in the language of graphs help us identify and focus on the essentials. The second portion takes up the Tower of Hanoi puzzle. Here, graphs convert a problem about finding an algorithm to finding a path in a diagram.

In Chapter 4, we move further towards algebra. We study the problem of sending a secret message which can be understood by the intended receiver, but not by anyone who manages to intercept it. The first step is to substitute numbers for the characters in the message. Once this is done, various algebraic tools become available for manipulating the numbers and scrambling the message. We take the reader through the entire process of converting a message into numbers and then encrypting and decrypting it through the clever use of algebra. The use of mathematical software is indispensable here, even for a short message.

Chapters 5 and 6 explore aspects of the geometric objects known as fractals. One of the realisations of the last half-century has been that many natural and economic phenomena involve 'self-similarity'—as we view these phenomena on different scales of space or time, we find the same patterns. This calls for a different perspective on geometry as intuition based on shapes likes circles and cones does not suffice. Chapter 5 introduces us to the simpler examples of self-similarity in geometry. Chapter 6 then illustrates how simple rules, on being applied repeatedly, tend to create fractal patterns. Here too, the role of technology is indispensable. This study could take full flight only with the development of computer graphics, as prior to that, it was impractical to develop images with useful resolution.

Symmetry is a property in which an object, whether physical or mathematical, remains unchanged under some operation. Identifying and classifying symmetries is an important task in all areas of mathematics and its applications. In Chapter 7, we look at symmetries of planar shapes. Technology plays its part in visualising them and in carrying out the operations such as rotation and reflection.

Finally, Chapter 8 shows how matrices can bring order to algebraic calculations. We first use matrices to study the problem of ranking teams in a tournament. Then we apply them to studying how different components of a population change with time, and the long-term effects of their interactions. The components could be by genetic constitution (genotypes) or by age. Many of our earlier themes converge in this chapter. Graphs are used to depict the results of tournaments. Symmetry enters the picture as we look for population distributions which would not change under the given interactions. When we try to figure out long-term trends we find iterative processes. And, technology makes it feasible for us to quickly explore the changes and identify the underlying patterns for which theoretical understanding should be developed.

Acknowledgements

Jonaki Ghosh would like to express her deep gratitude to her doctoral supervisors Professor J N Kapur (late) and Professor Kumkum Dewan for their invaluable guidance. She would like to thank Dr Shayama Chona for being a positive force in so many ways and for inspiring her to embark upon a career in education. Jonaki is immensely grateful to her husband Somonnoy and her daughter Antara, for their unconditional love and support. She is indebted to her father Dr Sudhansu S Bhattacharjee (late) and mother Banashri Bhattacharya who have been a pillar of strength during the journey of writing this book and to her sister Sanchari who never failed to inspire. She is thankful to her parents-in-law, Akhil Kumar Ghosh and Shikha Ghosh for always being there. She would also like to thank all her friends and associates for encouraging her and for believing in her.

Amber Habib remembers his grandfather, the late Abdul Qadeer, for showing him the logical development of basic mathematics through a set of finely graded problems. These problems offered both challenge and reassurance, and first caused the thought that a life spent with mathematics could be as satisfying as driving trains or flying aircraft!

Geetha Venkataraman would like to thank her doctoral supervisor Peter M Neumann (1940–2020). Much of her journey in mathematics was shaped and nurtured by him. The pandemic took Peter away just short of his eightieth birthday. He lives on in the mathematics he bestowed. She remembers his detailed and painstakingly written out corrections, every time she corrects a grammatical or mathematical error while reading the work of her research scholars. Geetha would also like to thank her partner Mahesh Rangarajan, her mother Visalakshi Venkataraman and her daughter Uttara Rangarajan for their unstinting support. Her father WgCdr P S Venkataraman (late) and her mother-in-law Shantha Rangarajan (late) were there in spirit to support this endeavour.

We thank Madhu Reddy, Director, Universities Press, for his encouragement of our proposal. We also appreciate the work put in by our editor, Gita Dattatri.

Jonaki Ghosh, Amber Habib, Geetha Venkataraman

Contents

1

Monte Carlo Methods: Estimating π

For centuries and generations, mathematicians have striven to estimate the value of π and to find more and more accurate approximations of it. Aryabhata worked out the value of π to five significant figures, that is, 3.1416. Archimedes had given a systematic method for approximating π. In his treatise *Measurement of a Circle*, it is stated 'The ratio of the circumference of any circle to its diameter is less than $3\frac{1}{7}$ but greater than $3\frac{10}{71}$'. He found these bounds by inscribing and circumscribing a circle with two similar 96 sided polygons. The number π continues to fascinate many.

In this chapter we will explore some unique and interesting methods of approximating the value of π. These are known as the Monte Carlo methods and they require the simulation of data by generating random numbers. Typically, any mathematical software has inbuilt random number generators and they provide opportunities to explore phenomena based on chance. The numbers generated by these random number generators are not actually random. They are generated as pseudo-random numbers requiring a sophisticated algorithm, inbuilt in the program. These pseudo-random numbers behave in ways similar to random numbers and can be used to study random phenomena. In this chapter we will first explore a method of estimating π, based on Monte Carlo simulation. This *Square–Circle method* is based on generating points randomly within a square of side two units with its centre at the origin.

For our second application of simulation, we shall study *Buffon's Needle Problem*, which was proposed in 1777 by Georges Louis Leclerc, Comte de Buffon, a French nobleman [41]. It is a well-known problem in geometric probability and is an application of using π in probability problems. We will take the help of `Mathematica`, a computer algebra system, and `MS Excel`, a spreadsheet program, to explore both methods in depth. `Mathematica` may be accessed online on Wolfram Cloud. The free software `LibreOffice Calc` may be used instead of `Excel`. Appendix A provides a basic introduction to the software used in this book.

The pre-requisite mathematical knowledge for this chapter is elementary calculus, probability and trigonometry.

1.1 Random Numbers

Our lives are full of events which have an element of unpredictability. We toss a coin and do not know whether heads will show up, or tails. We set out to visit a friend and cannot be sure if we will reach on time. We make an investment with our savings and are unable to predict just how much we will gain. Randomness, in examples like these, appears as an inconvenience and an obstacle to our plans. But it also has its uses. It is the randomness of a coin toss that makes it a fair technique for deciding who bats first in a cricket match. And a randomly selected bunch of numbers makes for a stronger PIN for our bank card.

Randomness, in general, is not utter chaos. It does have some patterns and if we can understand these patterns, we can deal with it better. How can we become more familiar with randomness so that we can gain insight into its patterns? One idea is that we try to mimic random behaviour through computers. A computer fundamentally deals with numbers, so what it can try to do is to produce strings of numbers which appear random. For example, it may produce a string of real numbers between 0 and 1 such as:

0.913, 0.195, 0.482, 0.355, 0.252, 0.415, 0.612, 0.841, 0.505, 0.678

These numbers can be considered random in that there is no obvious pattern as to how one number follows another. We would not be able to say which is the next number that the computer would produce.

We shall now describe some ways in which you can use a calculator or a computer to generate random numbers.

Fig. 1.1.1 A Casio scientific calculator with a random number generator. Image credit [C1].

Scientific calculators usually have functions called `RAN` or `RAND` that can generate random numbers. The calculator shown in Fig. 1.1.1 has a function marked `Ran#` which produces real numbers between 0 and 1, such as the ones given above.

If we wish to generate long sequences of random numbers and do more advanced work with them, it is more convenient to use a computer program. If you have `Microsoft Excel` installed on a computer, you can type `=RAND()` in a cell to generate a random real number between 0 and 1. `RANDBETWEEN(M,N)` generates integers between M and N. For example, if we type `RANDBETWEEN(1,6)` and then press the `enter` key, we will generate an integer between 1 and 6. This could be useful for exploring the possibilities in a game of dice.

	A	B
1	=RAND()	=RANDBETWEEN(1,6)
2	=RAND()	=RANDBETWEEN(-1,1)
3	=RAND()	=RANDBETWEEN(0,1)
4		
5		

	A	B
1	0.718533829	2
2	0.206928203	0
3	0.984613903	1
4		
5		

Fig. 1.1.2 `Excel` commands and resulting output

For more involved calculations, we may wish to move from `Excel` to a dedicated mathematics software such as `Mathematica`. In `Mathematica`, we can generate various kinds of random numbers through commands such as the following:

- `RandomReal[]`: generates a random real number between 0 and 1.
- `RandomReal[b]`: generates a random real number between 0 and b.
- `RandomReal[a,b]`: generates a random real number between a and b.
- `RandomReal[a,b,n]`: generates n random real numbers between a and b.
- `RandomInteger[]`: randomly generates 0 or 1.
- `RandomInteger[m,n]`: generates a random integer between m and n.
- `RandomInteger[m,n,p]`: generates p random integers between m and n.

Finally, several websites such as `www.random.org` also allow generation of random numbers.

Now that we have some idea of how random numbers can be generated, we need to also develop our understanding of their basic features. The various techniques given above have two common goals for the numbers they generate:

- The numbers should be **unpredictable**—even if we know many numbers of the sequence, we should be unable to predict the next one.
- The numbers should be **unbiased**—no particular number or range of numbers should be more likely than others within the allowed range.

The Figs 1.1.3 and 1.1.4 illustrate these points. They are based on a string of 1000 random numbers generated in `Excel`.

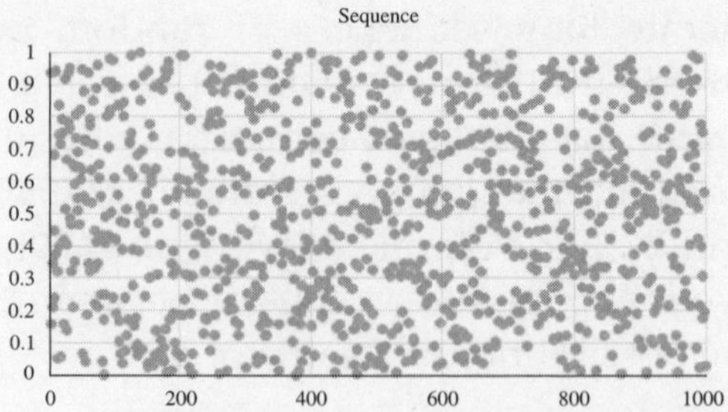

Fig. 1.1.3 This figure illustrates the *unpredictability* of the generated random numbers. Each random number is plotted as a circle. On the x-axis we have the location of the number in the sequence, from the first value to the thousandth. On the y-axis we have the value of the number. We see that the values jump around in a chaotic fashion.

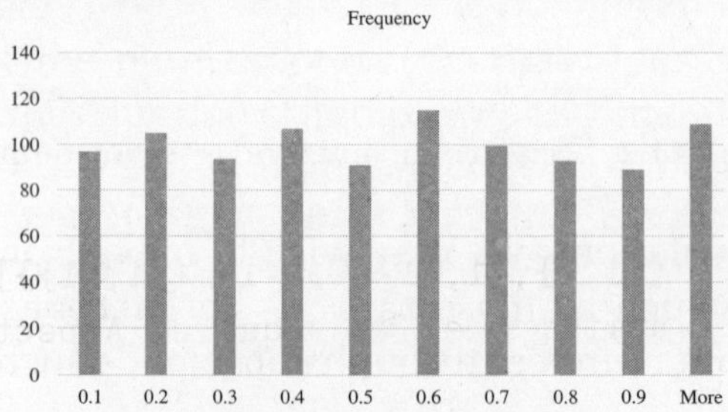

Fig. 1.1.4 This figure illustrates the *unbiased* nature of the generated random numbers. The chart shows how many numbers are between 0 and 0.1, between 0.1 and 0.2, and so on. We see that the random numbers are quite evenly distributed across the different parts of the 0 to 1 range.

1.2 THE SQUARE–CIRCLE METHOD FOR ESTIMATING π

Consider the unit circle with its centre at the origin, inscribed in a square of side 2 units (Fig. 1.2.1).

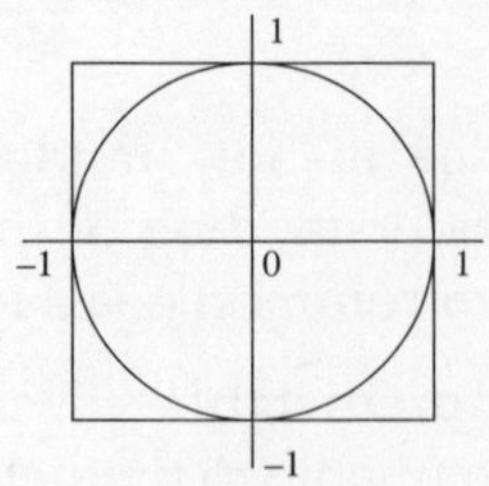

Fig. 1.2.1

Investigation 1.1 What are the coordinates of the vertices of the square?

For any point (x, y) within the square, we have $-1 \leq x \leq 1$ and $-1 \leq y \leq 1$. Also any point (x, y), which lies on or inside the circle satisfies the inequality

$x^2 + y^2 \leq 1$. However, those points (x, y) which lie outside the circle but inside the square, satisfy the inequality $x^2 + y^2 > 1$.

We shall now randomly generate 1000 ordered pairs (x, y) with both x and y lying between -1 and 1, using the following `Mathematica` code:

```
Table[{RandomReal[{-1, 1}], RandomReal[{-1, 1}]}, {i, 1, 1000}]
```

To plot the points generated by the above code within the square of side 2 and the unit circle we use the following program (the output of which is given in Fig. 1.2.2).

```
data=Table[{RandomReal[{-1, 1}], RandomReal[{-1,1}]}, {i,1,1000}];
plotpoints= ListPlot[data, PlotStyle->
 {PointSize[0.01], GrayLevel[0.2]}, AspectRatio->Automatic];
unitcircle= Graphics[{Thickness[0.01], Circle[{0,0},1]}];
unitsquare= Graphics[{Thickness[0.01],
 Line[{{-1, -1}, {-1,1},{1, 1},{1,-1},{-1,-1}}]}];
Show[plotpoints, unitcircle, unitsquare, AspectRatio->Automatic]
```

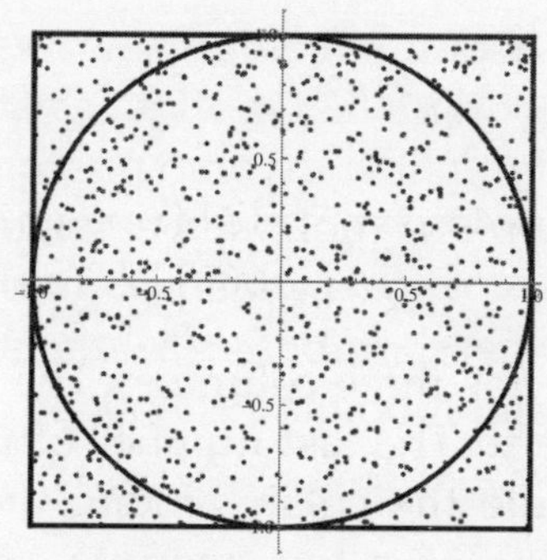

Fig. 1.2.2 `Mathematica` output for 1000 simulated points (x, y) where $-1 \leq x \leq 1$ and $-1 \leq y \leq 1$

Note that the 1000 points are fairly uniformly distributed within the square.

Investigation 1.2 Run the above code by replacing 1000 with 10000. What do you observe?

As we increase the number of points, they appear to fill the square more and more. How many of these points end up inside any region of the square? Intuitively, we expect that a larger region will attract more points. More precisely, we expect that the probability that a randomly selected point from the square will be in a particular region should be proportional to the area of the region. Now, the area of the square is 4 square units and the area of the circle is π square units. Hence the probability that a randomly selected point from the square will be inside the circle is $\pi/4$. Keeping in mind that probability reflects the relative frequency of a favourable outcome, we have the following approximation:

$$\frac{\text{Number of points lying inside the unit circle}}{\text{Total number of points inside the square}} \approx \frac{\pi}{4}$$

The above relation can be used to estimate the value of π. With 1000 points we may not get a good approximation. However, as the number of points is increased, this approximation gets better and better. This approach of estimating quantities with the help of random processes is called *Monte Carlo Simulation*.

The `Mathematica` code which we used to generate random points within the square does not count the number of points that were inside the circle. The following program rectifies this defect and counts all generated points (x, y) which satisfy the inequality $x^2 + y^2 \leq 1$. It generates two random numbers `r1` and `r2`, representing x and y respectively. The `If` condition checks whether a randomly generated ordered pair satisfies the condition $x^2 + y^2 \leq 1$. A counter called `points` gets incremented by 1 every time an ordered pair satisfies the required condition.

```
r1:=RandomReal[{-1, 1}];
r2:=RandomReal[{-1, 1}];
points=0;
Table[{random1=r1, random2=r2, If[(random1)^2+(random2)^2<=1,
 points+=1, outside]}, {I, 1, 1000}]//TableForm
```

The output of this program has three columns. The first two columns show the values of x and y respectively. The last number of the third column is the total number of points lying inside the circle, among the 1000 randomly generated points. To estimate the value of π we compute `4*points/1000`. The result is expressed as a fraction. To obtain it in decimal form, use `N[4*points/1000]`.

Investigation 1.3 Estimate the value of π using the above code for $n = 1000$. Run the code 10 times. Each time, note the number of points which lie inside the unit circle and enter your results in the table provided below. You now have 10,000 simulated points. Use this data to estimate the value of π. Is this approximation better than that obtained by using only 1000 throws?

	Number of points inside the circle (in 1000 throws)
1	
2	
3	
4	
5	
6	
7	
8	
9	
10	
Total (in 10,000 throws)	

Simulating the Problem in Excel

We can also simulate the problem in a spreadsheet program like `Excel`. Let us consider the quarter of the unit circle, which lies in the first quadrant. Its area is $\pi/4$. The area of the portion of the square in the first quadrant is 1 square unit.

We shall now use `Excel` to generate 100 points (x, y) randomly, with $0 \leq x, y \leq 1$, and count those which satisfy the condition $x^2 + y^2 \leq 1$. This can be done by using the `RAND` and `IF` functions. In column A we can simulate values of x, in column B values of y and in column C we can verify if the generated pair of values (x, y) satisfies the condition $x^2 + y^2 \leq 1$. The steps of simulation are as follows:

Step 1: The first step is to simulate 100 values of x in column A. Enter `=RAND()` in cell B2 and drag till cell B101. This function will generate a random number between 0 and 1 in each cell. 100 randomly generated real numbers will appear in column B. (When creating a column of numbers by dragging, we need to drag the small box in the lower right hand corner of the cell.)

Step 2: Simulate 100 values of y in column C, by first entering `=RAND()` in cell C2 and then double clicking in the corner of the cell.

Step 3: We now need to verify if the condition $x^2 + y^2 \leq 1$ is satisfied for each pair of values (x, y). This is done by entering `=IF(B2^2+C2^2<=1, 1, 0)` in cell `D2` followed by a double click on the corner of the cell. This step will produce 1 if the condition is satisfied and will give 0 if it is not satisfied. Thus column D comprises 0's and 1's only.

Step 4: To find the proportion of points lying inside the quarter circle we need to sum the 1's in column D and divide by 100. Enter `=SUM(D2:D101)/100` in any cell (say `G2`). To obtain the estimated value of π, we enter `=4*G2` in cell `G3`.

The output is shown in Fig. 1.2.3.

	A	B	C	D	E	F	G
1		x values	y values	Condition			
2	1	0.58822077	0.95581386	0			0.79
3	2	0.35539441	0.23940812	1		Estimate of Pi	3.16
4	3	0.93902361	0.87722512	0			
5	4	0.49210239	0.54525521	1			
6	5	0.69672022	0.29159823	1			
7	6	0.49877167	0.27686324	1			
8	7	0.4849527	0.57331269	1			
9	8	0.35714271	0.86286573	1			
10	9	0.76254199	0.97682485	0			
11	10	0.91688329	0.4377434	0			
12	11	0.23357881	0.06410027	1			
13	12	0.51071826	0.22288108	1			
14	13	0.55564962	0.26352914	1			
15	14	0.63554707	0.35663103	1			
16	15	0.47650061	0.82868912	1			
17	16	0.48540935	0.70388573	1			
18	17	0.67681097	0.07735669	1			
19	18	0.91188882	0.6819231	0			
20	19	0.69488759	0.07199093	1			

Fig. 1.2.3 An `Excel` simulation of the square–circle method of estimating π

Investigation 1.4 Run the simulation in `Excel` 10 times and estimate the value of π.

1.3 THE BUFFON'S NEEDLE PROBLEM

The Problem *Suppose you are given a sheet of paper with equidistant parallel lines drawn horizontally as shown in Fig. 1.3.1. Let us assume that the distance between each consecutive pair of lines is l. You are also given a thin needle of length k. Imagine that you throw the needle several times on the sheet of paper. What is the probability that the needle will fall in such a way that it will either touch a line or cross over in a particular throw?*

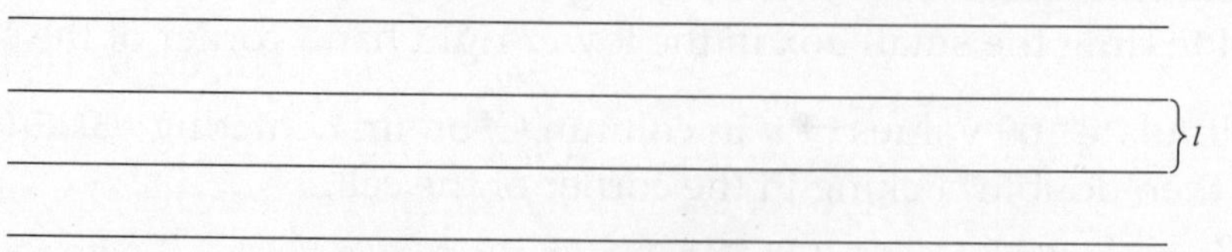

Fig. 1.3.1 Equally spaced parallel lines on a sheet of paper

In this section we shall explore this very interesting problem using both `Mathematica` and `Excel`. A similar exploration may be done with any other Computer Algebra System or spreadsheet program.

Let us first try to understand the problem. There are three possibilities for the length k of the needles, that is, $k < l$, $k = l$ and $k > l$. We will first discuss the case when $k = l$, that is, the needle length is equal to the distance between the parallel lines. When such a needle is thrown on the sheet of paper it could fall in several ways, some of which are indicated in Fig. 1.3.2. The case when the needle touches or crosses over a line will be referred to as a *hit*, whereas the case of landing between two lines without touching them will be referred to as a *miss*.

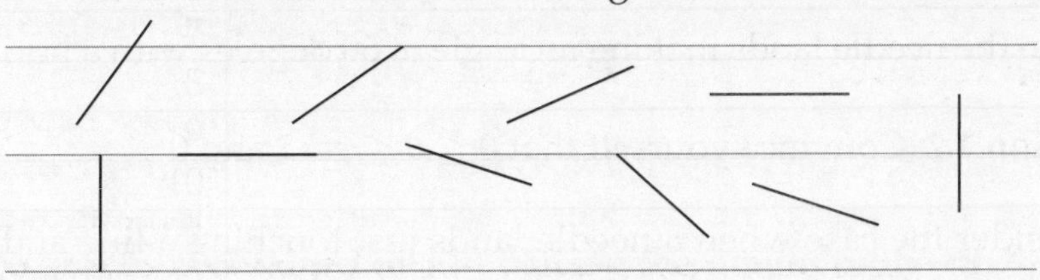

Fig. 1.3.2 Various needle throws

Investigation 1.5 Try to perform this experiment by throwing a needle on a sheet of paper with equally spaced lines about 50 times. What is the proportion of hits? Is there any pattern among the hits? How do we differentiate the hits from the misses?

Let us try to analyse the problem mathematically by defining two random variables x and θ as follows.

$$x \quad = \quad \text{distance of the centre of the needle from the nearest line,}$$
$$\theta \quad = \quad \text{the acute angle (in radians) made by the needle with the nearest line.}$$

If the needle lands parallel to the lines, we set $\theta = 0$.

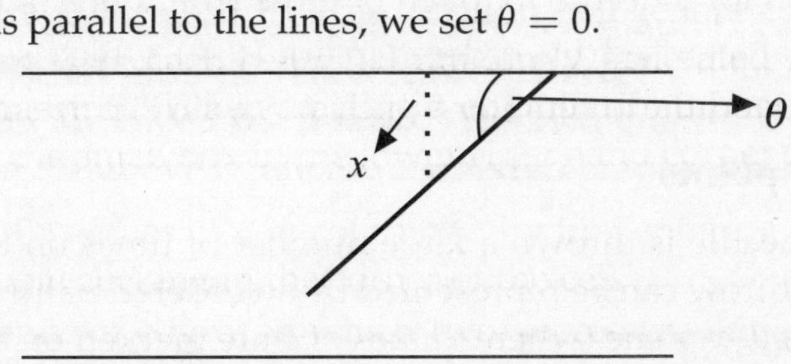

Fig. 1.3.3 Identifying θ and x for a needle throw

Every needle throw will have a unique value of x and θ and therefore can be identified by an ordered pair (θ, x).

Investigation 1.6 Can you identify the ordered pairs (θ, x) corresponding to the following needle throws?

(a) When the needle lands horizontally on top of a line?
(b) When the needle lands horizontally between two lines?
(c) When the needle lands vertically with its centre on a line?
(d) When the needle lands vertically exactly between two lines?
(e) When the needle lands making an angle of 60 degrees with a horizontal line?

Investigation 1.7 Convince yourself that $0 \leq \theta \leq \pi/2$ and $0 \leq x \leq l/2$.

Let us consider the case when a needle lands just touching a line and making an angle θ with the horizontal as shown in Fig. 1.3.4(a). It is easy to see that $\sin\theta = \frac{x}{l/2}$ or $x = (l/2)\sin\theta$. Figure 1.3.4(b), where the needle crosses over while making the same angle θ with the horizontal, indicates that $x < (l/2)\sin\theta$. Figure 1.3.4(c) shows that when the needle lands between two lines, then $x > (l/2)\sin\theta$.

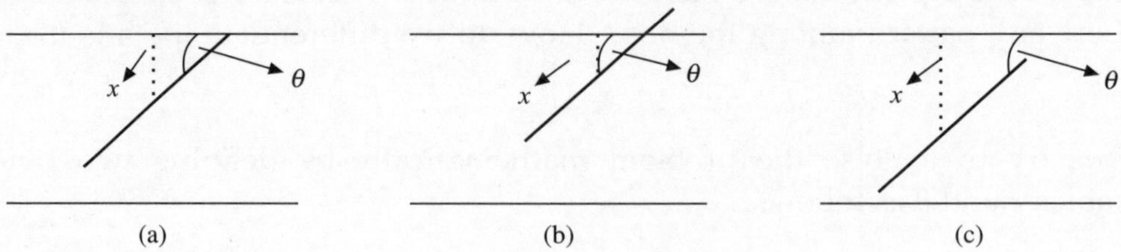

Fig. 1.3.4 Different needle positions

Thus, the condition for a needle to touch or cross over a line is $x \leq \frac{l}{2}\sin\theta$. Hence needle throws characterised by ordered pairs (θ, x) which satisfy $x \leq \frac{l}{2}\sin\theta$ correspond to *hits*, and the others (for which $x > \frac{l}{2}\sin\theta$) correspond to *misses*.

From needles to points

Imagine that the needle is thrown a large number of times on the sheet of paper. Since each needle throw can be represented by an ordered pair (θ, x), we can mark this as a point on a θ–x plane (where values of θ are marked on the horizontal axis and values of x on the vertical axis).

Clearly, the inequalities $0 \leq \theta \leq \pi/2$ and $0 \leq x \leq l/2$ imply that all possible values of (θ, x) will be contained within a rectangle with side lengths $\pi/2$ and $l/2$. Thus, for every needle throw we will obtain a point within this rectangle. We may refer to this rectangle as the *needle throw rectangle*.

The following Mathematica code may be used to visualise the needle throws as points within our needle throw rectangle.

```
l=1;
needlethrows=
 Table[{RandomReal[N[Pi/2]], RandomReal[l/2]}, {i, 1, 1000}];
needlepoints=
 ListPlot[needlethrows, PlotStyle->{PointSize[0.01],
 GrayLevel[0.4]}]
```

This is another instance of Monte Carlo simulation. It will enable us to approximate the desired probability of hits without actually performing the physical experiment of throwing needles on a sheet of paper.

In the above Mathematica code, the needle throws have been simulated using the Random command. Note that the code randomly generates 1000 pairs of values of θ and x and plots them using the ListPlot command. For simplicity, the distance l between the lines has been taken as 1 unit. The output is shown in Fig. 1.3.5, with θ plotted along the horizontal axis and x along the vertical axis.

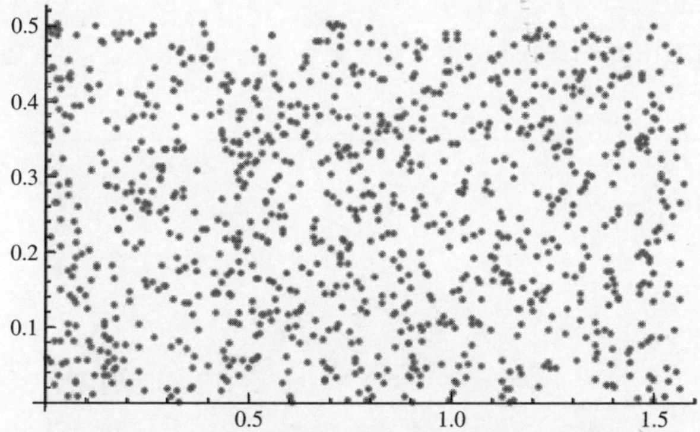

Fig. 1.3.5 Mathematica output showing 1000 needle throws as points within a rectangle of dimensions $\pi/2$ and $1/2$

We shall now draw the graph of the function $x = (l/2)\sin\theta$ within the rectangle. This may be done by adding the following lines to the above Mathematica code.

```
graph = Plot[(l/2)*Sin[x], {x, 0, Pi/2},
 PlotStyle->{Thickness[0.004], Black}, Filling->Bottom];
Show[needlepoints, graph]
```

Note that the *hits* will be represented by points on or below the graph of $x = (l/2)\sin\theta$. The *misses* will be points lying above the graph.

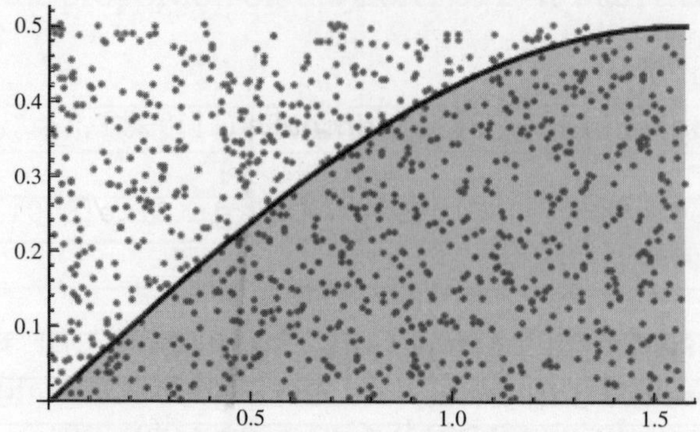

Fig. 1.3.6 `Mathematica` plot showing the graph of $x = (l/2)\sin\theta$ along with 1000 needle throws

Now, the probability p of a needle hitting (touching or crossing over) a line is the probability of a random (θ, x) point being below the graph $x = (l/2)\sin\theta$. Therefore,

$$p = \frac{\text{area under } x = (l/2)\sin\theta}{\text{area of the rectangle}}$$

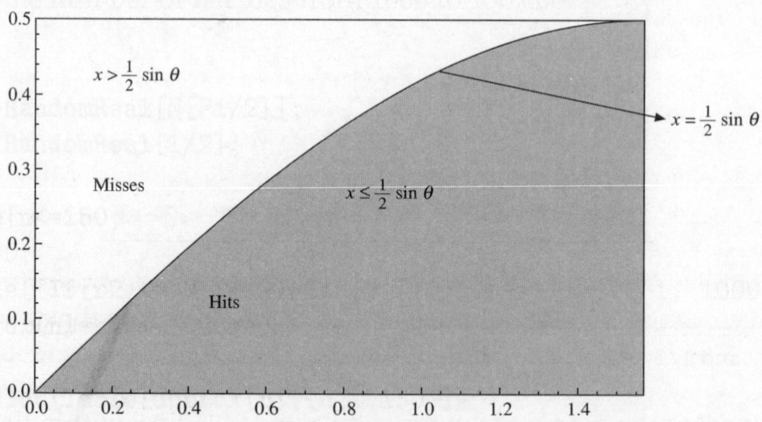

Fig. 1.3.7 The shaded area represents the proportion of hits

Investigation 1.8 Use integration to compute the shaded area under the graph of $x = (l/2)\sin\theta$ and show that the probability of a hit is $p = 2/\pi$.

Thus Buffon's problem can be solved theoretically, with the help of calculus. However, our goal is to use the problem to demonstrate how the Monte Carlo method can help us estimate probabilities *without* theory, in preparation for situations which are too complex to be easily resolved by theory.

If the number of throws is large enough, the proportion of hits will serve as an estimate of the corresponding probability. That is,

$$p = \frac{\text{shaded area}}{\text{area of the rectangle}} \approx \frac{\text{the number of hits}}{\text{total number of throws}}$$

Mathematica Code for Counting the Hits

Although the `Mathematica` code in the previous section helps us to understand the problem visually, it does not actually count the hits. The following code will do so. It generates two random numbers `r1` and `r2`, representing θ and x respectively. The `If` condition checks whether a randomly generated ordered pair, that is, a needle throw satisfies the condition $x \leq \frac{l}{2}\sin\theta$. A counter `hit` gets incremented every time an ordered pair satisfies the condition.

```
l=1;
r1:= RandomReal[N[Pi/2]];
r2:= RandomReal[l/2];
hit= 0;
Table[{random1=r1, random2=r2, If[random2 <= N[(l/2)*Sin[random1]],
 hit+=1, miss]}, {i, 1, 1000}]//TableForm
```

In the output of the code (Fig. 1.3.8). there are three columns.

```
0.208497   0.098889    1
0.478538   0.0327708   2
0.800934   0.404049    miss
0.190176   0.469992    miss
1.518      0.264401    3
```

Fig. 1.3.8 Counting hits

The first two columns show the values of θ and x respectively. The third column indicates the hits and misses. The last number of this column is the total number of hits.

Investigation 1.9 Run the above code 10 times. Each time note the number of hits and enter your results in the table provided below. You have 10,000 simulated needle throws. Is the proportion of hits closer to $2/\pi$ than that obtained by using only 1000 throws?

	Number of hits (in 1000 throws)
1	
2	
3	
4	
5	
6	
7	
8	
9	
10	
Total (in 10,000 throws)	

Naturally, we expect that throwing more needles will generate more accurate estimates of π. The following code runs the experiment 150 times, steadily increasing the number of needles from 1000 to 150,000.

```
l=1;
r1:= RandomReal[N[Pi/2]];
r2:= RandomReal[l/2];
n=1;
While[n<=150,
 hit=0;
 Table[{If[r2 <= N[(l/2)*Sin[r1]], hit+=1]},{i, 1, 1000*n}];
 approx[n]=hit/(1000*n);
 n++];
ListPlot[Table[approx[n],{n,1,150}]]
```

This gives the following output (see Fig. 1.3.9), confirming the increased accuracy as we throw more needles.

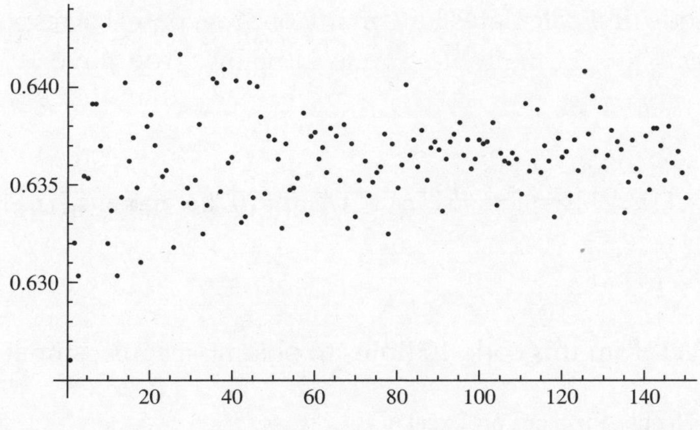

Fig. 1.3.9 Increasing the number of needles reduces the fluctuation in the probability estimates. The number of needles thrown (in thousands) is plotted horizontally, and the corresponding proportion of hits is plotted vertically.

Investigation 1.10 From Investigation 1.8 we know $p = 2/\pi$. Suppose the `Mathematica` simulations give us $p \approx 0.643$. Would we be be justified in using this to estimate π as follows: $\pi = 2/p \approx 2/0.643 = 3.11$?

The difficulty in combining our `Mathematica` code with Investigation 1.8 to estimate π, is that the code assumes the standard value of π when it generates random values of θ. We can get around this as follows. The basic principle is that acute angles correspond to ordered pairs in the first quadrant—each such ordered pair is joined to the origin and then we consider the angle between this line segment and the horizontal axis. To ensure that the angle selection is uniform, we only use the points which are inside the unit circle. Here is code which randomly selects 10,000 such pairs, calculates the corresponding slopes, and tabulates them with 10,000 randomly selected numbers between 0 and 1/2.

```
r1:= RandomReal[];
r2:= RandomReal[];
inside=0;
While[inside<=10000,
 random1=r1;
 random2=r2;
 If[random1^2+random2^2 <= 1, inside+=1;dir[inside]=random2/
 random1]];
throws=Table[{RandomReal[0.5],dir[n]},{n,1,10000}];
```

The following code first calculates how many of these pairs correspond to hits and then it estimates π.

```
hits=0;
For[n=1, n<=10000, n++,
 If[throws[[n,2]]>=throws[[n,1]]/Sqrt[0.25-throws[[n,1]]^2],
 hits+=1]];
pi=N[20000/hits]
```

Investigation 1.11 Run this code 10 times to obtain various estimates of π.

Simulation of the Needle Problem on Excel

The Monte Carlo method for Buffon's problem can be executed on `Excel` by using the functions `RAND`, `INT` and `IF`. In column A we can simulate values of θ, in column B values of x and in column C we can verify if the corresponding pair of values (θ, x) represents a hit or a miss. The steps of simulation are as follows.

Step 1: The first step is to simulate 100 values of θ in column A. Enter `=(Pi()/2)*RAND()` in cell `A2` and drag till cell `A101`. This will generate a random number between 0 and $\pi/2$ in each cell. The 100 randomly generated real numbers will appear in column A. (When creating a column of numbers by dragging, we need to drag the small box in the lower right hand corner of the cell.)

Step 2: The next step is to simulate 100 values of x in column B. Enter `=(0.5)*RAND()` in cell `B2` and double click in the corner of the cell. This will generate a random number between 0 and 1/2 in each cell.

Step 3: We now need to verify if the condition $x \leq (1/2)\sin\theta$ is satisfied for each pair of values (θ, x). This is done by entering `=IF(B2<=0.5*SIN(A2), 1, 0)` in cell `C2` followed by a double click on the corner of the cell. This step will produce 1 if the condition is satisfied (that is the throw is a hit) and 0 if it is a miss. Thus column C comprises 0's and 1's only.

Step 4: To find the proportion of hits we need to sum the 1's in column C and divide by 100. Enter `=SUM(C2:C101)/100` in any cell (say `E2`).

Investigation 1.12 Simulate 1000 needle throws in `Excel`. How does the proportion of hits compare to using 1000 throws generated by the `Mathematica` code?

	A	B	C	D	E	F	G
1	theta	distance (x)	hit/miss				
2	0.7814	0.2943628	1		0.64	3.125	
3	0.2526	0.0911729	1				
4	1.36956	0.0461619	1				
5	0.5519	0.3717675	0				
6	1.17878	0.1814626	1				
7	1.26857	0.1898914	1				
8	0.09269	0.3793411	0				
9	0.43038	0.2683337	0				
10	0.66892	0.1069973	1				
11	0.46703	0.343559	0				
12	1.30506	0.1856438	1				
13	1.33638	0.093202	1				
14	1.40727	0.2402734	1				
15	1.19445	0.4392843	1				
16	1.42782	0.3020456	1				
17	0.28379	0.1361241	1				
18	1.29502	0.482455	0				

Fig. 1.3.10 Simulating the Buffon's Needle problem using `Excel`

Extensions of the Needle Problem

What can we say about the probability if the length of the needle, k,

(a) Is less than the distance l between the lines?

(b) Is greater than the distance l between the lines?

These are explored in investigations 1.13 and 1.14.

Investigation 1.13 Compute the probability of the needle hitting a line if $k < l$.

Investigation 1.14 Compute the probability of the needle hitting a line if $k > l$.

Investigation 1.15 Suppose a coin of diameter d is thrown on a sheet of paper with horizontal and equidistant parallel lines (with a distance of 1 between each consecutive pair of lines). What is the probability that the coin will touch or cross over a line?

1.4 Chebyshev's Inequality and Accuracy of Estimates

Both in the Square–Circle method and in the Buffon's Needle problem, there were investigations which asked to compare the estimate of π when the number of points n or needles thrown were 1000 versus when the number is 10000. Intuitively, one would feel that the larger the n, the better the estimate of π. It is therefore natural to think of the following question: 'How large should n be so that we can estimate π accurately up to say 3 decimal places?' or equivalently, we want to know if there is a large enough n after which the estimate of π will always be 3.141.... We can find an answer to this and other related questions via a well-known inequality in Probability Theory called Chebyshev's Inequality. You can find out more about Chebyshev at [43].

Let us quickly remind ourselves about sample space, event, probability axioms and random variables. Our aim here is to be intuitive rather than very rigorous.

A *sample space* S is the set of all outcomes of an experiment whose outcomes are known but not predictable from before. An *event* E is a subset of the sample space. We say that an event E has occurred if the outcome of the experiment is in E. For example, if the experiment is a fair coin toss, then $S = \{H, T\}$, where H represents the outcome head and T tail. E can be any one of the four subsets of S. If the outcome of the toss is a head then the events $\{H\}$ and S have occurred but $\emptyset$ and $\{T\}$ have not occurred.

Suppose we are able to assign to each event E of a sample space S, a real number $P(E)$ such that the following are satisfied.

(i) $0 \leq P(E) \leq 1$;

(ii) $P(S) = 1$; and

(iii) If E_n is a sequence of events that are pairwise exclusive, that is, $E_n \cap E_m = \emptyset$ for $n \neq m$, then $P(\bigcup_{n=1}^{\infty} E_n) = \sum_{n=1}^{\infty} P(E_n)$.

Then P is called a *probability function* with the domain being the power set of S and the co-domain being the set of real numbers. $P(E)$ is called the *probability* that the event E occurs. Often we are not interested in the outcomes per se but with questions related to some functions of the outcomes. For example, when the experiment is of two fair dice being cast, the sample space is $S = \{(i, j) \mid i, j = 1, \ldots, 6\}$ and we might be interested in knowing the probability of the sum of the two numbers on the dice being 5, rather than the individual outcomes, or a question like 'what is the probability that the sum is larger than 6?'.

A *random variable* X is a function from the sample space S to the set of real numbers $\mathbb{R}$. A suitably defined random variable will help us pose our questions and answer them. In the example of the two fair dice being cast, we could define a random variable $X : S \to \mathbb{R}$ as $X((i, j)) = i + j$. So X denotes the sum of the numbers appearing on the two fair dice and our question can then be posed as 'what is $P(X > 6)$?'. In our example, X takes the values, $2, 3, 4, \ldots, 12$ and $P(X=5) = 4/36$ as it is the probability of the event $E = \{(1,4), (4,1), (2,3), (3,2)\}$.

Suppose that X is a random variable that takes the real value x_i with probability p_i, for $i = 1, \ldots n$. Then we define the *expected value* or *mean*, $\mu = E(X) = \sum_{i=1}^{n} x_i p_i$ and the *variance* $\sigma^2 = \text{var}(X) = E[(X - \mu)^2] = E(X^2) - \mu^2 = \sum_{i=1}^{n} x_i{}^2 p_i - \mu^2$. *Standard deviation* is given by σ.

Chebyshev's Inequality states that if X is a random variable with mean μ and variance σ^2, then for any positive real number $k > 0$,

$$P\{|X - \mu| \geq k\} \leq \frac{\sigma^2}{k^2}$$

If we were to choose $k = m\sigma$ then the above inequality states that the probability that X takes a value at a distance greater or equal to m times the standard deviation, away from the mean, is less than or equal to $1/m^2$. An alternative formulation of Chebyshev's Inequality which will help us in estimating π accurately is

$$P\{|X - \mu| < k\} > 1 - \frac{\sigma^2}{k^2}$$

But how does Chebyshev's Inequality help us in accurately estimating π? We will illustrate this for the Square–Circle method. In the Square–Circle method, we are throwing a point into the square of side two units, in which is inscribed the unit circle.

Let X_i be the random variable which takes the value 4 if the ith point thrown falls into the unit circle and the value 0 otherwise. We consider X_i taking the value 4, as success and X_i taking the value 0 as a failure. So the outcomes are divided into two categories, 'success' and 'failure'. Clearly, $P(X_i = 4) = \frac{\pi}{4} = p$ and $P(X_i = 0) = 1 - p$. Such a random variable is an example of a *Bernoulli* random variable. Clearly $E(X_i) = 4p = \pi$ and $\text{Var}(X_i) = 16p - \pi^2 = \pi(4 - \pi)$. You can find out more about Jacob Bernoulli at [40].

Now let $X = \frac{1}{n}\{X_1 + X_2 + \cdots X_n\}$. If $i \neq j$, the success or failure of the ith throw in no way influences that of the jth throw or vice-versa, we have that the X_i are independent random variables. So,

$$\mu = E(X) = \frac{1}{n}\{E(X_1) + E(X_2) + \cdots E(X_n)\} = \pi$$

and

$$\sigma^2 = \text{var}(X) = \frac{1}{n^2}\{\text{var}(X_1) + \text{var}(X_2) + \cdots \text{var}(X_n)\} = \frac{1}{n}\pi(4-\pi)$$

Using this in Chebyshev's Inequality, we get for any $k > 0$

$$P\{|X - \pi| < k\} > 1 - \frac{\{\frac{1}{n}\pi(4-\pi)\}^2}{k^2} = 1 - \frac{\pi^2(4-\pi)^2}{n^2k^2}$$

If we want to estimate π accurately to 3 decimal places, then we can choose $k = 0.0001 = \frac{1}{10000}$. The above equation then becomes

$$P\{|X - \pi| < 0.0001\} > 1 - \frac{(10000)^2\pi^2(4-\pi)^2}{n^2}$$

Suppose we want to predict with 95 percent accuracy or with greater than 0.95 probability, that the estimated value of π and the actual value of π, will differ by less than 1/10000, then we need to find a large enough n such that

$$1 - \frac{(10000)^2\pi^2(4-\pi)^2}{n^2} > 0.95$$

This gives the following inequality for n:

$$n > \sqrt{20} \times 10000 \times \pi(4-\pi)$$

Using $3 < \pi < 4$, we get that if $n > 50{,}000 \times 4 = 200{,}000$, then with 0.95 probability the estimated value of π will be accurate at least up to three decimal places to the actual value of π. So if we throw 200,000 points then we can with 95% probability, get the value of π correct to at least three decimal places. Of course, we have not tried to estimate the best possible n here but are only giving a rough idea. It is also clear that if we want, say, 99% probability, then we need an even larger n. Similarly, if we want accuracy up to 5 decimal places, then again we will require a much larger number of throws than 200,000.

Further Explorations

An interesting variation to Buffon's Needle problem, is *Buffon's Noodle Problem*. Here the needle, bent into a polygonal shape or replaced by a curve of unit length is randomly dropped on a plane marked with parallel lines one unit apart. The Wolfram Demonstration project site [79] shows an attractive simulation, while a mathematical treatment of the problem may be found in [59].

A more detailed introduction to Probability Theory and Chebyshev's Inequality can be found in [63].

2

Exploring Probability through Simulation

The topic of probability can be explored through many interesting games, which depend on chance. In this chapter we shall explore two problems. The first one is the well-known Monty Hall Problem, also known as the Three Doors Problem. The second one is a popular game called Craps, which is played in casinos. Both problems require the notion of conditional probability for their analysis and lend themselves to investigation. We shall explore the intricacies of these very interesting problems through simulation in `MS Excel` and also analyse them using conditional probability.

2.1 The Monty Hall Problem: A Historical Journey

The Monty Hall Problem is known by several names such as the Monty Hall Paradox, the Three Doors Problem, the Quiz-master Problem, and the Problem of the Car and the Goats. It was introduced by the biostatistician Steve Selvin way back in 1975 in a letter to the journal *The American Statistician* [64] and may be stated as follows:

> You are a contestant in a game show where the host (Monty) asks you to choose one among three doors. Behind one of the doors is a car (the prize) and behind the other two are goats. After you select a door, the host doesn't open it; instead, he opens one of the other two doors and reveals a goat (the host, who knows what's behind the doors, never reveals the car). The host then asks you whether you want to stick to your original choice or you would like to switch by choosing the other door. The question is, which option is more likely to win you the prize?

This problem and its solution was published by Marilyn vos Savant in her *Ask Marilyn* column in Parade magazine [78] after which the problem became very popular the world over. Marilyn claimed that the contestant should switch because the odds of winning the car would be 2/3, whereas if one sticks to the original door, then the odds of winning would be only 1/3. This solution resulted in angry retorts from many readers including professional mathematicians who argued that switching cannot matter since, after one of the goats is revealed by Monty, the probability of obtaining the car must be 1/2 as there are only two doors left. This problem has led to much controversy as Richard Gill [18, p. 2] remarks,

> The main question, which is endlessly discussed is whether simple arguments for switching, which typically show that the unconditional probability that the switching gets the car is 2/3, may be considered rigorous and complete solutions of MHP. The opposing view is that vos Savant's question is only properly answered by study of the conditional probability that the switching gets the car, given the initial choice of door by the player and door opened by the host.

Of course, the analysis of the problem can be done only after making the assumptions of randomness, that all three doors are equally likely to hide the car and further that Monty, the host, doesn't have a preference towards opening any of the doors.

A Preliminary Exploration of the Problem

The initial reaction of most people, when presented with the Monty Hall problem, is that switching does not matter and after Monty opens one of the doors and reveals a goat, the probability of winning the car is 1/2. However, this argument is incomplete. The probability that the player will initially select the door with the car is 1/3, and that a wrong door is selected is 2/3. Thus, if the player's initial choice was correct then switching would lead him to lose. While, if the player chose a wrong door initially (the probability of which is 2/3) then switching would lead to a win (since the host will reveal the goat behind the other wrong door). These cases have to be incorporated into our analysis.

Let us now try to analyse the problem using a tree diagram (see Fig. 2.1.1). The first level of branches indicate that the probability of the location of the car being behind any one of the three doors is 1/3. Further, the player is equally likely to choose any one of the doors and the probability of each choice is 1/3. This is indicated by the second level of branches. Finally, Monty will open a door depending on the player's choice. Thus if the car is behind door 1 and the player has chosen the same door, then Monty can open door 2 or door 3 with probability

1/2 each. However, if the car is behind door 1 and the player has selected door 2, then Monty has no choice but to open door 3 (that is, the probability of his opening door 3 is equal to 1).

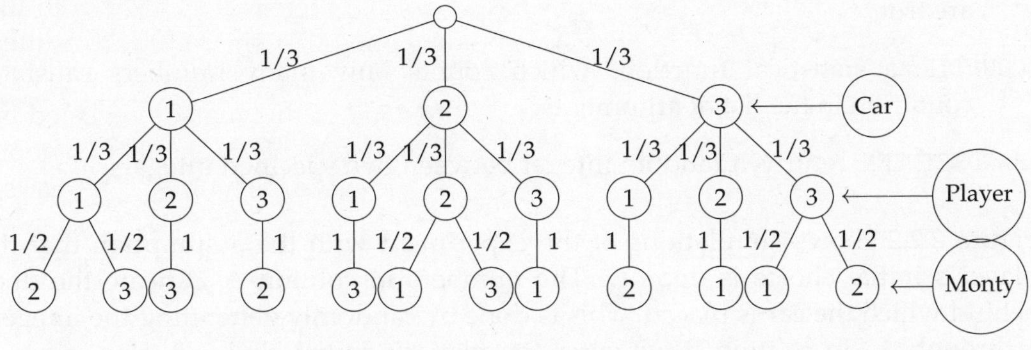

Fig. 2.1.1 A tree diagram representing the various outcomes of the Monty Hall Problem

Using the tree diagram in Fig. 2.1.1 we can work out the various outcomes of the game. We shall list the elements of the sample space in the form of triples (c, p, m) where c denotes position of the car, p is the player's choice and m is the door opened by Monty. Further, the probability of a particular outcome can be worked out by multiplying the probabilities along the branches of the tree leading to that outcome. For example, the outcome $(1, 1, 3)$ can occur with a probability $\frac{1}{3} \times \frac{1}{3} \times \frac{1}{2} = \frac{1}{18}$, whereas the probability of the occurrence of the outcome $(2, 1, 3)$ is $\frac{1}{3} \times \frac{1}{3} \times 1 = \frac{1}{9}$. Observe that the outcomes are not equally likely.

Investigation 2.1 List all the outcomes of the sample space and work out their respective probabilities. Using this information can you conclude that the probability of winning by switching is 2/3 whereas the probability of winning by staying with one's original choice is 1/3?

2.2 Spreadsheet Simulation of the Monty Hall Problem

In this section we shall simulate the Monty Hall problem in `MS Excel` by generating random numbers. In order to perform the simulation we need to be familiar with the following basic functions of `Excel`.

`IF`: a logical function, which specifies a logical test to be performed or a condition to be checked.

`OR`: a logical function, which returns `TRUE` if *any* of the arguments of a statement is true.

`AND`: a logical function, which returns `TRUE` if *all* of the arguments of a statement are true.

`COUNTIF`: a statistical function, which counts how many numbers satisfy a condition in the list of arguments.

`RANDBETWEEN`: returns a random integer between two specified integers.

Figure 2.2.2 shows simulations of the experiment with the assumption that the player's initial choice is door 1. The numbers in column A indicate the door behind which the car is placed. This is done by randomly generating the integers 1 through 3. In column B we enter the player's initial choice, that is, door 1. Column C indicates the door opened by Monty, the host. To simulate this, we check whether the corresponding entries in columns A and B are equal. If these are identical (that is, the player has selected the door with the car, in this case door 1) then Monty can open either door 2 or door 3 with equal probability (he cannot open the door chosen by the player). To simulate this, we generate a random number between 0 and 1. A random number less than 0.5 indicates Monty's choice of door 2 in which case 2 is entered in column C. However, if the random number is 0.5 or greater, then 3 is entered in column C indicating Monty's choice of door 3. These entries in column C are filled using a simple conditional statement.

In column D we identify the cases where the player wins by switching. If the entries of columns A and B are different, then `YES` appears in the corresponding cell (since the player's choice and location of the car are different). Similarly in column E we identify the cases where the player wins by staying with his original choice. Thus if the entries of columns A and B are the same then `YES` appears in the corresponding cell. Finally, counting the `YES`'s in column D reveals the total number of times the player wins by switching in 100 trials. Interestingly, this number, on an average turns out to be around 67. Thus one may conclude that the probability of winning by switching is 67% whereas the probability of winning by staying with one's choice is around 33%.

The steps of the simulation are as follows.

Step 1: The first step is to simulate the location of the car behind one of the three doors. This is done by generating the integers 1, 2 or 3 in column A, 100 times. This can be done by entering `=RANDBETWEEN(1,3)` in cell `A2` and dragging till cell `A101`. 100 randomly generated integers (1, 2 or 3) will appear in column

A. These represent the door behind which the car is placed in 100 different simulations of the game.

A	B	C	D	E
location of Car	Player's Choice	Door opened by Monty	Win by Switching from original choice	Win by Staying with original choice
=RANDBETWEEN(1,3)				

Fig. 2.2.1 The RANDBETWEEN command simulates the location of the car

Step 2: In this simulation, we are assuming that the player's choice is door 1. Enter 1 in cell `B2` and drag till `B100` or simply double click on the corner of cell `B2`. Note that all entries in column B will be 1.

Step 3: To simulate the door opened by Monty, enter `=IF(A2=1,IF(RAND()<0.5,2,3), IF(A2=2,3,2))` in cell `C2`, followed by a double click on the corner of the cell. This step simulates the door opened by Monty. Note that if the car is behind door 1 and the random number generated by `RAND()` is less than 0.5 then Monty opens door 2, otherwise he opens door 3. On the other hand if the car is behind door 2, then Monty has no option but to open door 3. Similarly if the car is behind door 3, Monty has to open door 2.

A	B	C	D	E
location of Car	Player's Choice	Door opened by Monty	Win by Switching from original choice	Win by Staying with original choice
3	1	2		
2	1	3		
1	1	3		
3	1	2		
3	1	2		
1	1	2		
1	1	3		
3	1	2		
1	1	3		
2	1	3		

Fig. 2.2.2 Simulating the Monty Hall problem using Excel

Step 4: In column D we identify the cases in which the player wins by switching and in column E we identify the cases where the player wins by staying with his original choice. For each row, the output of the cell in column D should be `YES` if the player's choice is not the same as the position of the car and should be `NO` otherwise. This is achieved by entering `=IF(A2<>B2,"YES","NO")` in cell `D2` followed by a double click in the corner of the same cell. Similarly, the output of the cell in column E should be `YES` if the player's choice is the same as the position of the case and NO otherwise. This can be done by entering =IF(A2=B2,"YES","NO") in E2 followed by a double click in the corner of the cell.

A	B	C	D	E
location of Car	Player's Original Choice	Door opened by Monty	Win by Switching from original choice	Win by Staying with original choice
1	1	2	No	No
2	1	3	Yes	No
1	1	2	No	No
1	1	2	No	No
3	1	2	Yes	No
2	1	3	Yes	No
1	1	3	No	No
1	1	2	No	No
3	1	2	Yes	No
3	1	2	Yes	No

Fig. 2.2.3 Simulating the Monty Hall Problem using Excel

Step 5: Finally, we need to count the number of YES's in columns D and E to find the number of cases in which the player wins by switching or by staying with his choice respectively. This can be obtained by entering =COUNTIF(D2:D101, "YES")/100 and =COUNTIF(E2:E101,"YES")/100 respectively in separate cells.

A	B	C	D	E	F	G	H
location of Car	Player's Original Choice	Door opened by Monty	Win by Switching from original choice	Win by Staying with original choice		Probability of winning by switching	Probability of winning by staying
3	1	2	Yes	No			
3	1	2	Yes	No		0.66	0.34
3	1	2	Yes	No			
1	1	3	No	Yes			
2	1	3	Yes	No			
1	1	2	No	Yes			
2	1	3	Yes	No			
2	1	3	Yes	No			
1	1	2	No	Yes			
3	1	2	Yes	No			

Fig. 2.2.4 Simulating the probabilities of winning by switching or staying with one's choice

Investigation 2.2 Modify the simulation by entering the player's choice as door 2. Does this have any impact on the probability of winning by switching or staying with one's choice?

Investigation 2.3 Modify the simulation by varying the probability of Monty's choice of opening door 2 or door 3 (when the player has chosen door 1). Let us assume that Monty has a preference for door 3 and opens it 3/4 of the time. What impact does this have on the probability of winning by switching or staying?

2.3 The Monty Hall Problem and Conditional Probability

In this section we shall use the notion of conditional probability to analyse the Monty Hall problem and find the theoretical probability of winning by switching or staying with one's choice.

Let C_i denote the event that the car is located behind door i. This implies $P(C_i) = 1/3$ where $i = 1,2,3$. Also let M_{ij} denote the event that Monty opens door j when the player chooses door i. Here i denotes the door chosen by the player and j denotes the door opened by Monty.

Now, using the notation of conditional probability, we have $P(M_{ij}|C_k) = 0$ if $i = j$, since Monty will never open the door chosen by the player. (Note that in this notation k indicates the location of the car). Also $P(M_{ij}|C_k) = 0$ if $j = k$, since Monty will never open the door which has the car behind it. If the player chooses the door with the car behind it (that is, $i = k$), then Monty can open any of the other two doors with probability 1/2 each, thus $P(M_{ij}|C_k) = 1/2$ for $i = k \neq j$. For the case $i \neq k$, that is, the case when the player chooses a door which does not have the car behind it, then Monty has no option but to open the third door (the one which was neither chosen by the player nor has the car behind it). Hence $P(M_{ij}|C_k) = 1$ for $j \neq i,k$. Overall, we have:

$$P(M_{ij}|C_k) = \begin{cases} 0, & j = i \text{ or } j = k \\ 1/2, & j \neq i = k \\ 1, & i,j,k \text{ all distinct} \end{cases}$$

Now let us compute the probability of the event M_{23}, that is, the event that Monty opens door 3 when the player chooses door 2. Using conditional probability we have,

$$\begin{aligned} P(M_{23}) &= P(M_{23} \cap C_1) + P(M_{23} \cap C_2) + P(M_{23} \cap C_3) \\ &= P(M_{23}|C_1)P(C_1) + P(M_{23}|C_2)P(C_2) + P(M_{23}|C_3)P(C_3) \\ &= 1 \times \frac{1}{3} + \frac{1}{2} \times \frac{1}{3} + 0 \times \frac{1}{3} = \frac{1}{2} \end{aligned}$$

We would now like to compute the probability of the event $C_1|M_{23}$, that is, the event that the car is actually behind door 1 given that Monty opens door 3 when the player chooses door 2. This would give us the probability of the player winning the car by switching from his original choice. Similarly, we would like to compute the probability of the event $C_2|M_{23}$, that is, the event that the car is actually behind door 2 given that the player chose door 2 and Monty opened door 3. This would give us the probability of the player winning the car by staying with his original choice. The computations are as follows

$$P(C_1|M_{23}) = \frac{P(M_{23} \cap C_1)}{P(M_{23})} = \frac{P(M_{23}|C_1)P(C_1)}{P(M_{23})} = \frac{1 \times (1/3)}{1/2} = \frac{2}{3}$$

$$P(C_2|M_{23}) = \frac{P(M_{23} \cap C_2)}{P(M_{23})} = \frac{P(M_{23}|C_2)P(C_2)}{P(M_{23})} = \frac{(1/2) \times (1/3)}{1/2} = \frac{1}{3}$$

Thus $P(C_1|M_{23}) = 2/3$ indicates that the probability of the player winning the car by switching from his original choice is 2/3 and $P(C_2|M_{23}) = 1/3$ indicates that the probability of the player winning the car by staying with his original choice is 1/3.

An Extension

We shall now try to extend the problem by assuming that there are four doors instead of three. The probability of the car being behind any one of the four doors is 1/4. Similarly, the player may choose any one of the four doors with a probability of 1/4 each. If the player's choice is the same as the position of the car, then Monty can open any one of the remaining three doors with probability 1/3 each. Thus, if the player has chosen door 1 and the car is also behind door 1, then Monty can choose to open one of the doors 2, 3 or 4 with probability 1/3 each. However, if the player's choice and the position of the car are different, then Monty has a choice of the remaining two doors (since he cannot reveal the car nor open the door chosen by the player), and he can open any one of them with probability 1/2.

Investigation 2.4 Draw a tree diagram to represent the four-door case. List out the outcomes of the sample space and evaluate the probability of winning by switching and winning by staying with one's choice. What can you conclude?

Investigation 2.5 Design an `Excel` simulation for the four-door case.

Investigation 2.6 What can you say about the probabilities of winning by switching or staying with one's choice if there are 5 doors? Can you extend the idea to n doors?

A Variant of the Monty Hall Problem

There are many variants to the Monty Hall Problem. One of the more popular ones was proposed by Martin Gardner and it goes as follows:

Three prisoners X, Y and Z have been sentenced for a crime they have committed and have been incarcerated in three separate cells in the prison. The judge has decided to randomly select one of the prisoners and set him free! A prison guard has overheard a conversation between the judge and the warden and knows which prisoner will be freed. However he cannot disclose this to the prisoners for fear of losing his job. The three get to know about this and are anxious to know which one of them is the chosen one. They know that the probability of any one of them being set free is 1/3. Prisoner X, the cleverest among them, decides to try his luck by finding out more information. Knowing that the guard will not reveal the name of the fortunate prisoner, he tells the guard to whisper the name of one (among Y and Z), who is *not* going to be set free. Also, he adds, that if he (that is, X) is to be freed, the guard may toss a coin to decide whom to name among Y and Z.

The guard is mystified by this suggestion but feels there is no harm in revealing the name of a prisoner who is not going to be set free. So he tells X that Z is *not* going to be freed. On hearing this, Prisoner X feels happy that his chances of getting free has increased from 1/3 to 1/2 (since, according to him, either he or Y is going to be free) and shares the secret with Y. However, Y is of a different opinion. He believes that X's chances of being free are still 1/3, but his own chances of freedom have gone up to 2/3! Which prisoner is correct? It is up to the reader to draw the analogy with the Monty Hall Problem!

2.4 The Game of Craps

In this section we shall explore a very popular game played in casinos. It is called the Game of Craps, or just Craps, and is played simply by rolling a pair of dice. It is said that if you want your money doubled quickly based on a game of pure chance, then the game of Craps is a great choice since the probability of winning this game is very close to 1/2. We will analyse this game by simulating it on `MS Excel`.

Rules of the Game of Craps

To play the game we roll a pair of dice, which leads to 36 possible outcomes. The sum of the numbers on the two faces can range from 2 to 12. If 1 appears on both the dice then the sum is 2 while the sum is 12 if 6 appears on both dice. It is also possible to obtain all the intermediate sums 3, 4, 5, 6, 7, 8, 9, 10 and 11.

Investigation 2.7 Compute the probability of each sum from 2 to 12.

Let us now familiarise ourselves with the rules of the game. A player rolls a pair of dice leading to three possible outcomes:

1. A total of 7 or 11 is obtained on the first roll. In this case, the player wins the game.
2. A total of 2, 3 or 12 is obtained on the first roll. In this case the player loses the game.
3. A total of 4, 5, 6, 8, 9 or 10 is obtained on the first roll. In this case the player neither wins nor loses (in the first roll) but can roll again. The total obtained becomes the player's *point* and the game continues, that is, the player rolls the dice again. Two situations may arise:
   (a) The player's point appears before a total of 7 in subsequent rolls. In this case the player wins the game.
   (b) A total of 7 appears before the point in subsequent rolls. In this case, the player loses the game.

For example, let us suppose that the player's point is 4, that is, a sum of 4 appears in the first roll. To continue the game he rolls the dice till he gets 4 (his point) or 7. Note that other sums are inconsequential and may be ignored. If 7 comes up before 4 the player loses whereas if 4 comes before 7, the player wins.

Probability of Winning the Game of Craps

Now we shall compute the theoretical probability of winning a single game of Craps. The player wins if outcomes 1 or 3(a), described above, take place. The outcomes leading to the sums 7 or 11 are (1,6), (2,5), (3,4), (4,3), (5,2), (6,1), (5,6) and (6,5) respectively. Thus 8 out of 36 possible outcomes are favourable for a win on the first roll leading to a probability of 8/36 or 22.22%.

Next, we calculate the probability of winning after getting a total of 4, 5, 6, 8, 9 or 10 (which becomes the player's point) in the first roll. In order to find the probability of winning by getting a total of, say 6, we need to find the probability of getting 6 in the first roll and then getting 6 before 7 in subsequent rolls. This is a case of conditional probability and it can be calculated as follows.

The probability of getting a 6 in the first roll is 5/36. In subsequent rolls, we need to consider only totals of 6 and 7. Other totals may be ignored. This leads to only 11 possibilities: (1,5), (2,4), (3,3), (4,2), (5,1) being the possibilities for a total of 6 along with the six possibilities for a total of 7. Among these, a win is possible only when a total of 6 precedes a total of 7 (in subsequent rolls after the first roll), the probability for which is 5/11. Thus the total probability of winning by getting a 6 in the first roll is $\frac{5}{36} \times \frac{5}{11} = 0.063$. Similarly, we can calculate the probability of winning by getting a total of 4, 5, 8, 9 or 10 in the first roll.

Investigation 2.8 Compute the theoretical probability of winning the game of Craps by getting a total of 4, 5, 8, 9 or 10 in the first roll. Hence compute the total probability of winning a game of Craps.

2.5 Spreadsheet Simulation of the Game of Craps

In this section we shall try to analyse the game through a simulation in `Excel`. In addition to the `RANDBETWEEN`, `AND`, `OR`, and `IF` commands used in the simulation of the Monty Hall problem, we will also use the `SUM` command which is a function that adds up all the values of an argument.

The steps of the simulation are as follows:

Step 1: Enter the numbers 1 to 20 in column A to indicate the number of rolls on the pair of dice. (Enter 1 in cell `A2` and `=A2+1` in cell `A3`. Drag cell `A3` till `A21`.)

Step 2: To simulate the numbers appearing on a pair of dice we enter `=RANDBETWEEN(1,6)` in `B2` and `C2` as shown in Fig. 2.5.1.

A	B	C	D
	Die 1	Die 2	Sum
1	=RANDBETWEEN(1,6)		
2			
3			

Fig. 2.5.1 The `RANDBETWEEN` command in `Excel` randomly generates an integer between 1 and 6

Step 3: After this we select the cells `B2` and `C2` and double click in the corner of cell `C2`. This will fill both columns with numbers from 1 to 6 simulating 20 rolls of a pair of dice as shown in Fig. 2.5.2. Now enter `=SUM(B2,C2)` or `=B2+C2` in cell `D2` and double click in the corner of cell `D2`. The first entry in column D will indicate if the player has won, lost, or if the game is to be continued.

A	B	C	D
	Die 1	Die 2	Sum
1	6	1	7
2	2	3	5
3	2	3	5

Fig. 2.5.2 Use of `Excel` to simulate rolls of a pair of dice. The entry in cell `D2` determines if the player wins, loses, or if the game will continue

Step 4: The conditional statement =IF(OR(D2=7,D2=11),1,0) in cell E2 allows us to see the outcome of the first roll (as shown in Fig. 2.5.3). The value of cell E3 will be 1 if the player obtains a sum of 7 or 11, that is, if the player wins, and is 0 otherwise.

A	B	C	D	E	F
	Die 1	Die 2	Sum		
1	6	1	7	=IF(OR(D2=7,D2=11),1,0)	
2	2	3	5		
3	2	3	5		

Fig. 2.5.3 The IF conditional statement is used to check if the player wins in the first roll

Similarly the conditional statement =IF(OR(D2=2,D2=3,D2=12),1, 0) in cell F2 (as shown in Fig. 2.5.4) leads to the output 1 in cell F2 if the sum on the first roll is 2, 3 or 12, that is, if the player loses and is 0 otherwise.

A	B	C	D	E	F	G	H
	Die 1	Die 2	Sum				
1	2	2	4	0	=IF(OR(D2=2,D2=3,D2=12),1,0)		
2	6	2	8				
3	1	4	5				

Fig. 2.5.4 The IF conditional statement is used to check if the player loses in the first roll

Step 5: Finally we enter the conditional statement =IF(AND(E2=0,F2=0),‘‘YES’’, ‘‘NO’’) in cell G2. The output of this statement (YES or NO) indicates if the game will continue beyond the first roll. Note that if the game does not continue beyond the first roll, we will have to ignore the remaining rows of the output.

A	B	C	D	E	F	G	H	I
	Die 1	Die 2	Sum					
1	5	1	6	0	0	=IF(AND(E2=0,F2=0),"Yes","No")		
2	2	4	6					
3	6	2	8					

Fig. 2.5.5 The IF conditional statement is used to check if the game continues beyond the first roll

Step 6: If the game continues beyond the first roll, we need to insert the following conditional statements in cells E3, F3 and G3 as follows and evaluate them by dragging till row 21.

1. In cell E3 enter =IF(OR(G2=‘‘NO’’,G2=‘‘’’),‘‘’’,IF(D3=$D$2,1,0))
2. In cell F3 enter =IF(OR(G2=‘‘NO’’,G2=‘‘’’),‘‘’’,IF(D3=7,1,0))
3. In cell G3 enter =IF(OR(G2=‘‘NO’’,G2=‘‘’’),‘‘’’,IF(AND(E3=0,F3=0),‘‘YES’’, ‘‘NO’’))

A	B	C	D	E	F	G	H
	Die 1	Die 2	Sum				
1	2	6	8	0	0	Yes	
2	2	6	8	=IF(OR(G2="No",G2=""),"",IF(D3=$D$2,1,0))			
3	5	3	8	IF(logical_test, [value_if_true], [value_if_false])			

Fig. 2.5.6

A	B	C	D	E	F	G	H
	Die 1	Die 2	Sum				
1	3	6	9	0	0	Yes	
2	4	4	8	0	=IF(OR(G2="No",G2=""),"",IF(D3=7,1,0))		
3	4	6	10				

Fig. 2.5.7

D	E	F	G	H	I	J	K
Sum							
9	0	0	Yes				
8	0	0	=IF(OR(G2="No",G2=""),"",IF(AND(E3=0,F3=0),"Yes","No"))				
4			IF(logical_test, [value_if_true], [value_if_false])				

Fig. 2.5.8 The `IF` condition is used to compute the cells `E3`, `F3` and `G3` to continue the game further

Step 7: Once the `IF` conditional statements have been entered in the cells `E3`, `F3` and `G3`, the game can be simulated by clicking on any cell of column B or C which represent the die rolls. Each click will result in a new simulation.

Step 8: Finally we would like the player to quickly see (by looking at the final output) if he has won or lost. As we have seen, a win can happen on the first roll (by getting a 7 or 11) or by getting a point (4, 5, 6, 8, 9, 10). Similarly the player may lose on the first roll (by getting 2, 3 or 12) or by getting a point. The following conditional statements will display the result.

1. In cell `J3` enter `=IF(OR(D3=7,D3=11),"YES","")`
2. In cell `J5` enter `=IF(OR(D3=2,D3=3,D3=12),"YES","")`
3. In cell `J7` enter `=IF(SUM(E4:E23)=1,"YES","")`
4. In cell `J9` enter `=IF(SUM(F4:F23)=1,"YES","")`

Investigation 2.9 Analyse the game by exploring different simulations:

(a) The player wins the game in the first roll (by getting a 7 or 11).

(b) The player wins the game by getting a total of 4, 5, 6, 8, 9 or 10 in the first roll.

The reader may wish to refer to Wolfram Mathworld [81] for a simple explanation of the probability of winning a game of Craps.

FURTHER EXPLORATIONS

It would be worthwhile to go through [38] for a detailed analysis of the Monty Hall problem and its variants. The MHP also has applications in game theory, where the problem is treated as a finite two stage, two person zero sum game. Here the two players are the contestant and the host. An online simulation of the MHP is available on the website [34] where the player can play the game by clicking on the 'Play' button. A detailed analysis of the Game of Craps may be found in [71]. There are many games of chance, which may be simulated on a spreadsheet. The Gambler's Ruin problem and the Problem of Points described in detail in The Blog on Probability and Statistics [36] may lead to interesting investigations.

3

Graph Theory: Seven Bridges Problem and the Tower of Hanoi

3.1 A Puzzle with Seven Bridges

In the 1700s, the city of Königsberg was part of East Prussia. The river Pregel flowed through the city and created an island called Kniephof. In 1736, the famed mathematician Leonhard Euler published a paper [12] in which he described and solved a problem that had become popular among the citizens of Königsberg. Here is a translation of Euler's introduction to the problem [39]:

> 'The problem, which I understand is quite well known, is stated as follows: in the town of Königsberg in Prussia, there is an island A called the "Kneiphof"; with the two branches of the river (Pregel) flowing around it, as shown in Fig. 3.1.1 (Figure 1 from Euler's paper). There are seven bridges, a, b, c, d, e, f and g, crossing the two branches. The question is whether a person can plan a walk in such a way that he will cross each of these bridges once but not more than once.'

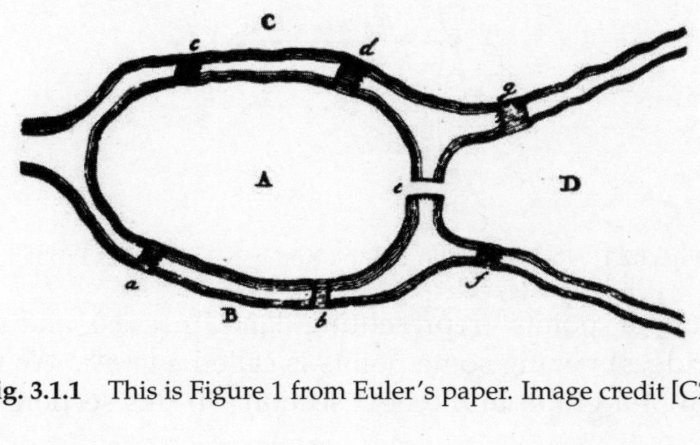

Fig. 3.1.1 This is Figure 1 from Euler's paper. Image credit [C2].

See [19] for an analysis of Euler's solution and [46] for more information about Euler.

The local names for the bridges were Grün (green), Köttel (guts, giblets), Krämer (shopkeeper), Schmied (blacksmith), Honig (honey), Höhe (high) and Holz (wood, wooden) respectively. Königsberg is now the city of Kaliningrad in Russia and after World War II and modernisation, in 2000, there were 5 bridges on the river Pregel [72].

Euler's paper is often cited as leading to the birth of Graph Theory. Not only is Graph Theory important as a branch of mathematics, it is important to computer science and in many applications in the real world. However, Euler himself did not mention or use graphs in his original paper in solving the Königsberg Bridges problem. The problem itself, as stated by Euler, is the following: *Is there a walk starting from one of the four land masses, A, B, C or D and ending in one of them, covering all the seven bridges a, b, c, d, e, f and g such that each bridge is crossed only once*? Euler showed that the answer to this problem was no and also gave a method to find out the solution to similar problems.

3.2 The Königsberg Bridges Problem and Graphs

Notice that we can represent Fig. 3.1.1 in an easier manner. For describing the problem of crossing the bridges, the shape of the land mass or the bridges do not matter. So we can shrink the land masses to points and the bridges can be lines or curves joining the points. If we do this we get the following representation for Fig. 3.1.1.

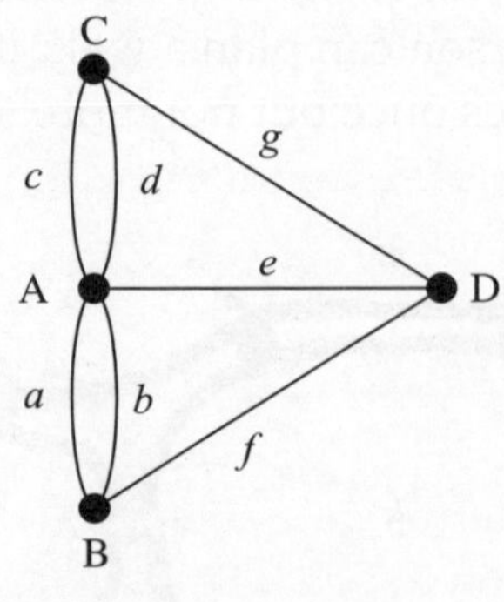

Fig. 3.2.1 Representation of the Königsberg Bridges Problem

Such a collection of points (representing land masses) and lines or curves (representing bridges) joining some points is called a *graph*. We will give a more formal definition of a graph in the next section. In this section, we will explore the conversion of Euler's Königsberg Bridges Problem to a problem of Graph Theory. We will also explore this kind of graph traversing problem via simple and maybe familiar examples. The aim is to arrive at an intuitive understanding of the solution that Euler presented in graph theoretic terms.

We shall call the points in a graph as *vertices* and the lines or curves joining the points as *edges*. So what was the problem posed by Euler, when we consider the graph in Fig. 3.2.1? Simply stated, starting from any of the vertices A, B, C or D can we trace all the edges without lifting the pen and without travelling the same edge twice? Does this remind you of something you may have tried when you were younger. Consider the graph below which has four vertices and ten edges.

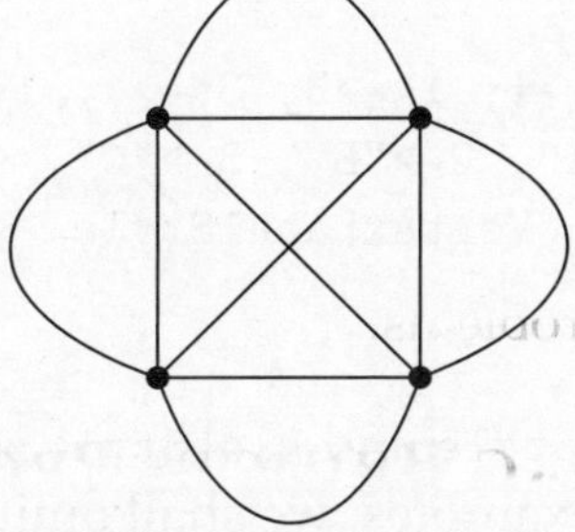

Fig. 3.2.2 A familiar graph

This is one of the popular graphs which is given to children with the problem being 'Can you draw this picture without lifting your pencil and without retracing any line?' After several attempts one soon realises that either an edge remains not drawn or one has to retrace an edge to be able to draw this graph without lifting the pencil. In other words, without resorting to trickery or cheating, the answer to the question is 'No'.

There are several questions we could explore, namely, is this a problem that can be posed for graphs in general? Are there graphs where the answer to such a problem is yes? If so, can we decide when the answer will be yes and when will it be no? Are there other variations to this problem and can we decide what the answer will be?

To explore these further, let us gather some evidence via investigations. But, before that, a brief introduction to some `Mathematica` commands that can be used to draw and investigate graphs will be useful. The `Mathematica` code

```
GraphPlot[{1<->2, 1<->2, 1<->3, 3<->1, 1<->4, 2<->4, 3<->4}]
```

gives the following unlabelled version of the KBP graph as output.

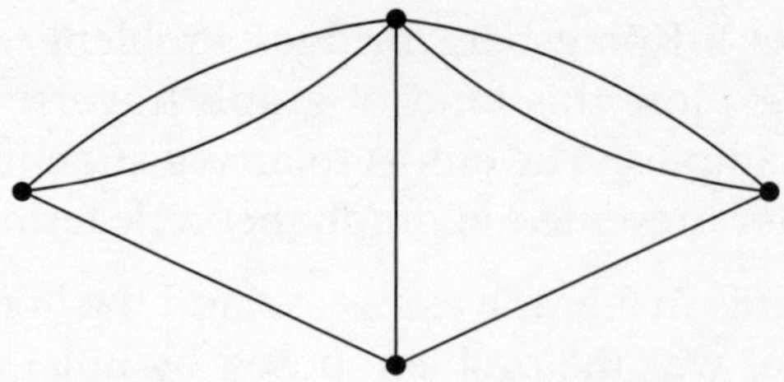

Fig. 3.2.3 Unlabelled graph produced by `GraphPlot`

Note that the part of the code `{1<->2, 1<->2,...}` indicates that two distinct edges are to be drawn between vertices 1 and 2. Here is another command to draw the same graph, but the vertices have also been labelled:

```
GraphPlot[{1<->2, 2<->1, 1<->3, 3<->1, 1<->4, 2<->4, 3<->4},
 VertexLabels->{1->"A", 2->"B", 3->"C", 4->"D"},
 EdgeStyle ->{Thick}, VertexLabelStyle ->Directive[Large],
 PlotStyle->Gray]
```

The `EdgeStyle`, `VertexLabelStyle` and `PlotStyle` settings control the formatting of the graph. In subsequent examples, we shall omit these settings for brevity. We get the following labelled graph as output:

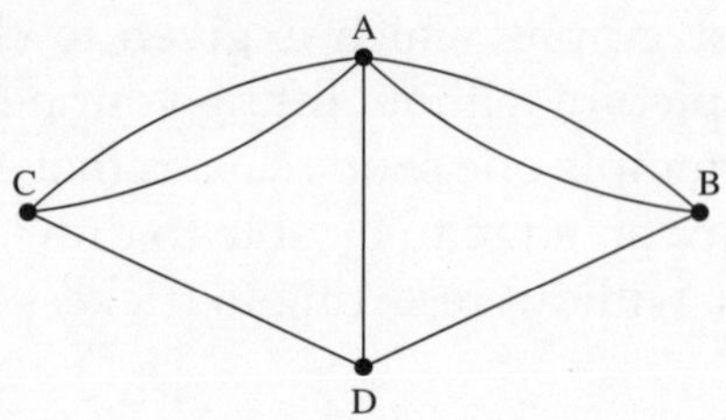

Fig. 3.2.4 Labelled graph produced by `GraphPlot`

One can also automatically label n vertices of a graph sequentially from 1 to n by the command

```
GraphPlot[{1<->2, 2<->1, 1<->3, 3<->1, 1<->4, 2<->4, 3<->4},
 VertexLabels -> "Name"].
```

The output is given below.

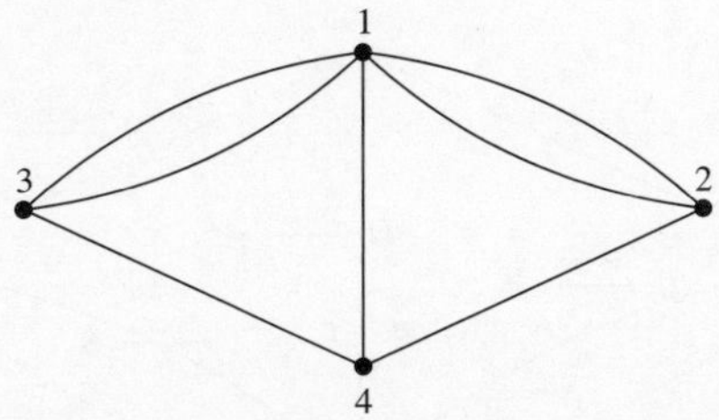

Fig. 3.2.5 Labelling vertices of the KBP graph automatically

However, these commands do not help us differentiate the multiple edges that are there between the same pair of vertices. To draw the KBP graph with the labelling as in Fig. 3.2.1 we use the following commands.

```
EdgeTaggedGraph[{A<->B, B<->A, A<->C, C<->A, D<->A, B<->D, C<->D}->
 {"a","b","c","d","e","f","g"}, VertexLabels->"Name",
 EdgeLabels->"Name"]
```

This gives the output shown below.

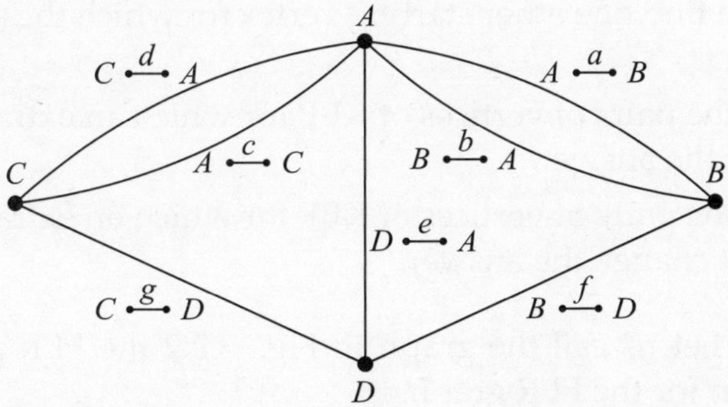

Fig. 3.2.6 KBP graph with vertices and edges labelled

Now suppose that we want to trace a particular walk on the KBP graph, say `A->B`, `B->D`, `D->A`, `A->C`, `C->A`. Let's say we want to label the edges as FIRST, SECOND and so on to describe the route. Then the following commands may be used. Note that only the edges involved in the walk are being shown as directed edges (with arrows) so that the route followed is clear.

```
EdgeTaggedGraph[{A->B, B<->A, A->C, C->A, D->A, B->D, C<->D}->
 {"FIRST", "x", "FOURTH", "FIFTH", "THIRD", "SECOND", "y"},
 VertexLabels->"Name", EdgeLabels->"Name"]
```

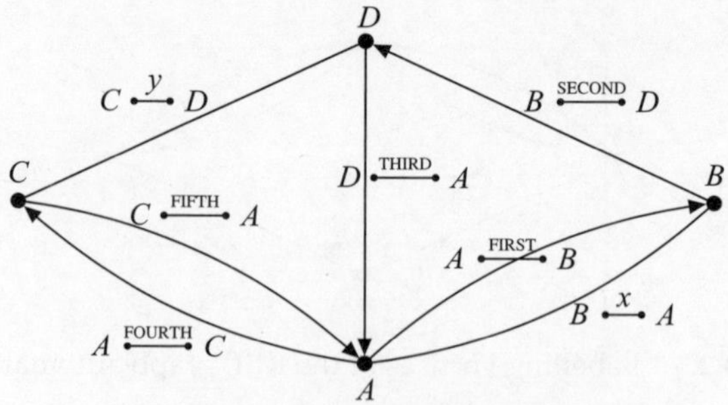

Fig. 3.2.7 KBP graph with a walk

Investigation 3.1 Let us call the graph in Fig. 3.2.1 the KBP graph. The investigations below may be carried out using the `Mathematica` commands discussed above as well as manually.

(a) Add an extra edge between one pair of vertices of KBP and decide whether the answer to the 'Bridges problem' changes from no to yes.

(b) If the answer changed to yes, note the starting vertex and end vertex for a successful tracing of the edges without lifting the pen or retracing.

(c) If you change the starting vertex, does the answer still remain yes? If not, see if you can find one other starting vertex for which the answer still remain yes.

(d) Find out all the pairs of vertices of KBP for which an extra edge between the pair changes the answer.

(e) Find out all the pairs of vertices of KBP for which an extra edge between the pair does not change the answer.

Investigation 3.2 Let us call the graph in Fig. 3.2.2 the FLR graph. Repeat the above investigation for the FLR graph.

We define the *degree* of a vertex of a graph to be the number of edges emanating or joining or incident to the vertex. Let us examine the KBP graph. We see vertex A has degree 5 as the edges a, b, c, d, e are incident to A. Examining the graph at other vertices, we see that B, C and D have degree 3.

`Mathematica` can also be used to find the degrees of all vertices in a given graph and also of individual vertices. As a first step the graph in question is specified. First, we create a graph and give it the name 'g':

```
g = GraphPlot[{1<->2, 2<->3, 3<->4, 4<->1, 1<->3, 2<->4, 1<->3},
 VertexLabels->"Name"]
```

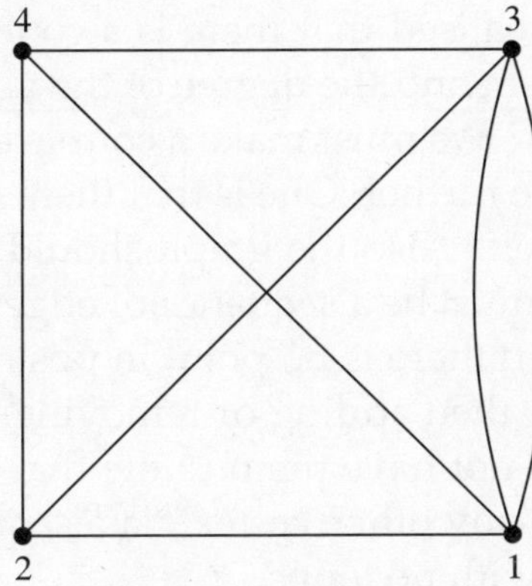

Fig. 3.2.8 Graph 'g'

The command `VertexDegree[g]` returns the output $\{4, 3, 4, 3\}$ which tells us that Vertex 1 has degree 4, Vertex 2 has degree 3, Vertex 3 has degree 4 and Vertex 4 has degree 3. The command `VertexDegree[g, 2]` gives output $\{3\}$, namely the vertex degree of Vertex 2.

Investigation 3.3 Find the degrees of the vertices of the FLR graph. Do so manually and using `Mathematica`.

Investigation 3.4 Do the following by using `Mathematica` and manually.

(a) If by adding an extra edge to a pair of vertices of the KBP graph, the answer to the Bridges problem changed to yes, then examine the degrees of all the vertices for the corresponding modified graph.

(b) If by adding an extra edge to a pair of vertices of the FLR graph, the answer to the Bridges problem changed to yes, then examine the degrees of all the vertices for the corresponding modified graph.

Investigation 3.5 In both the KBP graph and FLR we found that all the vertices had an odd degree. Now try the following:

(a) Draw a graph with all vertices of even degree.

(b) Draw a graph with exactly two vertices of even degree and all others of odd degree.

(c) Can you draw a graph in which there are an odd number of vertices with odd degree?

Investigation 3.6 For the graphs drawn in the above investigation, what is the answer you get for the Bridges problem?

By now you must have realised that there is a connection between a positive answer to the Bridges problem and the degree of the vertices of a graph. Can you guess what the connection is? We must make a couple of basic assumptions about our graphs before we explore further. One is that there should be no loops, that is, an edge from a vertex to itself. Also the graph should be *connected*, that is, if we pick a pair of vertices, there must be a sequence of edges taking one to the other. If a graph is not connected then there is no point in posing the Bridges problem for it. If the graph is connected, then adding or removing loops does not change the answer, so we might as well not have them. Note that in a connected graph, you can walk from one vertex to any other vertex via a sequence of edges. The figure below is a connected graph with no loops.

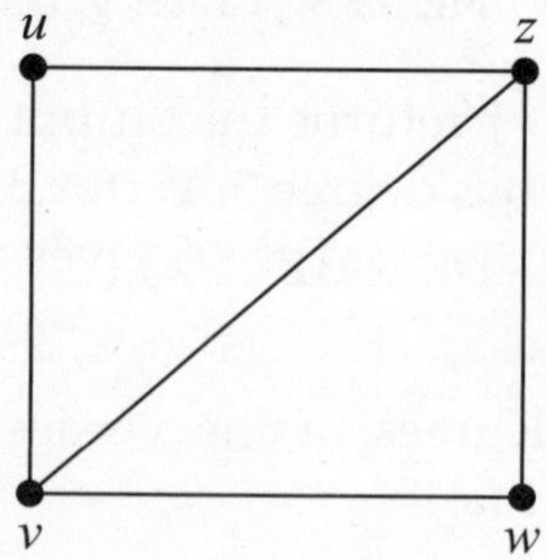

Fig. 3.2.9 A connected graph

Let us examine a graph for which we have a positive answer to the Bridges problem. This means that there is a vertex in the graph where we can begin, trace every edge without lifting the pen and without repeating an edge, and then end at another vertex. Let us call the starting vertex A and the end vertex Z. Other vertices will be called 'in-between' vertices. Think of this tracing the edges without repeating as a walk.

Let C be an in-between vertex. Since C is an in-between vertex, and the walk has to cover all the edges exactly once, the walk will pass through C as many times as there are edges incident to C. Since we neither started at C, nor we end at C, for every edge via which we enter C, there has to be another by which we exit it. Since no edge can be used twice in our walk, we can pair the edges incident to C as 'enter and exit' edges. So, the degree of C has to be even. In fact this argument shows that if we have a successful walk or a positive solution to the Bridges problem, then all in-between vertices in the walk must have even degree.

What about the degrees of A and Z? Since we start at A, there is a starting edge. If A has other edges, then as we do not end at A, we must be able to pair these other edges as entry and exit edges. Thus degree of A will be odd. Similarly Z will also have an odd degree as we have an edge, which we shall call an end edge, for finishing at Z and if there are more edges, they must exist as entry and exit pairs.

We have by the above analysis, actually come up with the result given below.

If the Bridges problem is solvable for a given connected graph, then the graph has exactly two vertices of odd degree and all other vertices have even degree.

Now we can pose another question, namely, whether the converse of the above result is true. The answer actually is yes but to prove it we will need more machinery and the formal setting of graph theory. This converse, which is true, is given below.

Let G be a connected graph, which has exactly two vertices of odd degree and the rest of even degree. Then we can start at one of the odd degree vertices and trace the graph without lifting the pen such that all the edges get traced exactly once and we will end at the other vertex of odd degree.

There is also a related question we can ask: *Is it possible to start at a vertex of a connected graph, trace all the edges exactly once and end at the starting vertex*? We shall examine this and the connected Bridges problem in the next section in a more formal setting.

3.3 Graphs, Walks, Trails and Paths

We will begin with the formal definition of a *simple graph*. A simple graph G consists of two sets:

(i) A non-empty finite set $V(G)$ of elements called vertices (or nodes).

(ii) A finite set $E(G)$ of distinct unordered pairs of distinct elements of $V(G)$ called edges.

The set $V(G)$ is called the *vertex set* of G and the set $E(G)$ is called the *edge set* of G. Basically, $E(G)$ consists of some two element subsets of $V(G)$. When G is fixed, we shall use V and E instead of $V(G)$ and $E(G)$.

An *edge* $\{v, w\}$ is said to join the *vertices* v and w, and is usually abbreviated to vw. Here $vw = wv$. By definition, in a simple graph we cannot have loops, that is, edges from a vertex to itself. We also cannot have multiple edges between two vertices. It is possible for $E(G)$ to be an empty set. KBP is not a simple graph as it has some pairs of vertices with multiple edges between them.

Below are some figures of simple graphs.

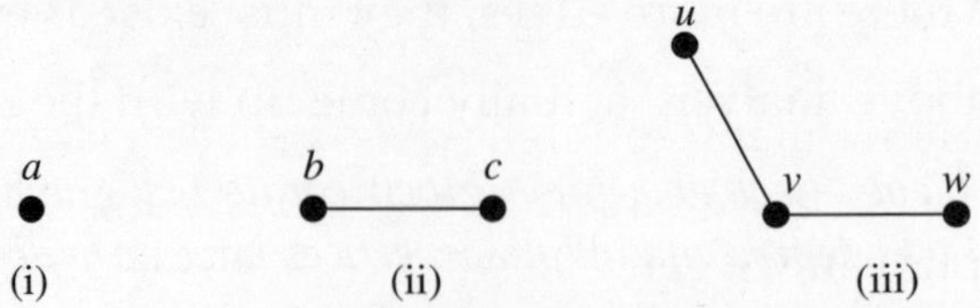

Fig. 3.3.1 Simple graphs

For the simple graph in Fig. 3.3.1 (i), we have $V = \{a\}, E = \emptyset$. For the graph in Fig. 3.3.1 (ii), we have $V = \{b, c\}$ and $E = \{bc\}$. For the third graph (iii) in the same figure $V = \{u, v, w\}$ and $E = \{uv, vw\}$.

Investigation 3.7 Draw the graph in Fig. 3.2.9 using `Mathematica`. Is FLR a simple graph? What about the graph in Fig. 3.2.9? If either of these is a simple graph, then identify the vertex set and the edge set.

Investigation 3.8 Draw all possible simple graphs on three vertices by hand and using `Mathematica`. Find the vertex set and edge set for each of the graphs.

We now define a *graph G*. It consists of two sets:

(i) A non-empty finite set $V(G)$ of elements called vertices (or nodes).

(ii) A finite family $E(G)$ of unordered pairs (of not necessarily distinct) elements of $V(G)$ called edges.

All simple graphs are graphs but a graph need not be a simple graph. Consider the graph in Fig. 3.3.2. It is clearly not a simple graph as the vertex z has a loop and the vertices u, w have multiple edges. For the graph G in Fig. 3.3.2, the vertex set $V = V(G) = \{u, v, w, z\}$ and the edge set $E = E(G)$ consists of 6 edges, namely uv, vw, uw, uw, wz and zz. So in a graph we allow loops and multiple edges between two vertices.

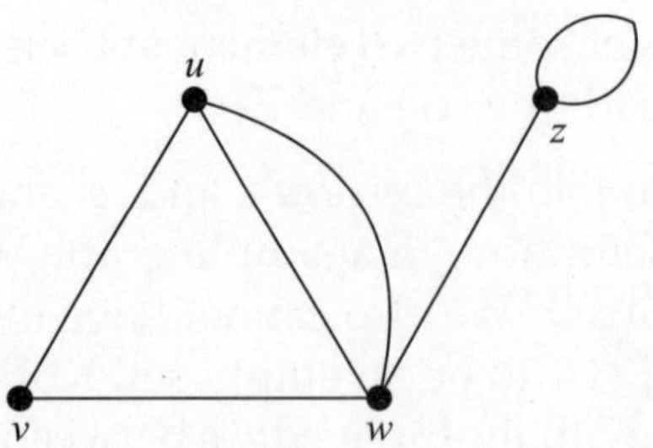

Fig. 3.3.2 Graph with a loop and multiple edges

Investigation 3.9 By removing edges and/or loops convert the graph in Fig. 3.3.2 into a simple graph.

Investigation 3.10 How many different graphs G can you draw with $V(G) = \{u, v\}$ and $|E(G)| = 3$? If we have no condition on the size of $E(G)$ then how many different graphs can we get?

Next we explore the *union* of two graphs and *connectedness*. Let $G_1 = (V_1, E_1)$ and $G_2 = (V_2, E_2)$ be two graphs where V_i denotes the vertex set of G_i and E_i denotes the edge set of G_i respectively for $i = 1, 2$. Let $V_1 \cap V_2 = \emptyset$. We denote the *union* of the two graphs by $G_1 \cup G_2$ and it is defined as the graph with vertex set $V_1 \cup V_2$ and edge set $E_1 \cup E_2$. Since the vertex sets are disjoint, the figure for the union of two graphs is obtained by just placing the figures of the two graphs next to each other. In the example displayed below, G_1 is the graph on the left with vertices $\{u, w, z\}$ and G_2 is the graph on the right with vertices $\{a, b, c\}$. Figure 3.3.3 represents $G_1 \cup G_2$.

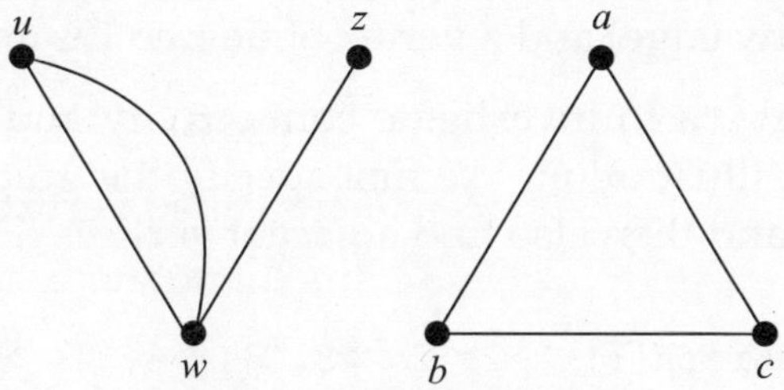

Fig. 3.3.3 Union of graphs

A graph is *connected* if it cannot be expressed as the union of two graphs, and disconnected otherwise. Clearly, any disconnected graph G can be expressed as the union of connected graphs. The connected graphs whose union gives the disconnected graph are called *components*. The two graphs, G_1 (on the left) and G_2 (on the right) in the above figure are connected graphs whereas $G_1 \cup G_2$ is not connected.

Note that, a graph is connected if and only if between any two vertices a and b, there is a sequence of edges, in which two successive edges have a common vertex, the first edge is incident at or begins at a and the last edge ends at b. In Fig. 3.3.2, consider the vertices v and z. The edge vw, wz form a sequence of edges as described above. Another such sequence is vu, uw, wz.

Investigation 3.11 Check whether KBP, FLR and the graphs in Fig. 3.3.1 and Fig. 3.3.2 are connected graphs. If they are connected, can you disconnect them by creating a new graph by removing some edges or vertices? If so, what is the least number of vertices or edges required to disconnect the respective graphs?

Some more definitions for a graph G.

Adjacent vertices: Two vertices v and w are adjacent if there is an edge vw joining them.

Adjacent edges: Two distinct edges e and f are adjacent if they have a vertex in common, that is, $e = vw$ and $f = wx$ for some vertices v, w, x.

In the KBP graph, vertices B and C are not adjacent, whereas A is adjacent to B, C and D; C is adjacent to A and D; and D is adjacent to A, B, C. Also in KBP, any two edges are adjacent. In Fig. 3.2.9 the edges wz and uv are not adjacent.

Degree of a vertex: The degree of a vertex v is the number of edges incident with v, and is written as deg v; in calculating the degree of v, we make a rule that a loop at v contributes 2 (rather than 1) to the degree of v. A vertex of degree 0 is an isolated vertex (not joined to any edge) and a vertex of degree 1 is an end-vertex.

Let us use `Mathematica` to investigate connectivity and adjacency of vertices through the following illustration. We first specify the graph for which we want to verify connectivity and then also find adjacent vertices. First,

```
gr = EdgeTaggedGraph[{1<->2, 2<->3, 3<->4, 4<->1, 1<->3, 2<->4,
1<->3, 5<->6, 6<->7, 7<->5, 8<->8}, VertexLabels ->"Name",
 EdgeLabels ->"Name", EdgeStyle ->{Thick}]
```

gives us the graph in Fig. 3.3.4 with vertices and edges labelled. Also note that in the case of multiple edges, the labelling accounts for the same.

The command `ConnectedGraphQ[gr]` asks if the graph 'gr' is connected and the output we get is 'False'. We follow this with the command `ConnectedComponents[gr]` to get the output $\{\{2,1,3,4\},\{6,5,7\},\{8\}\}$ which indicates that there are three components, and each is a connected graph. If we denote them by $\mathtt{gr}_1, \mathtt{gr}_2$ and $\mathtt{gr}_3$ respectively, then we see that $\mathtt{gr} = \mathtt{gr}_1 \cup \mathtt{gr}_2 \cup \mathtt{gr}_3$.

If the previous command is followed by `HighlightGraph[gr,%]` then the vertices in the respective connected components get highlighted with the same colour.

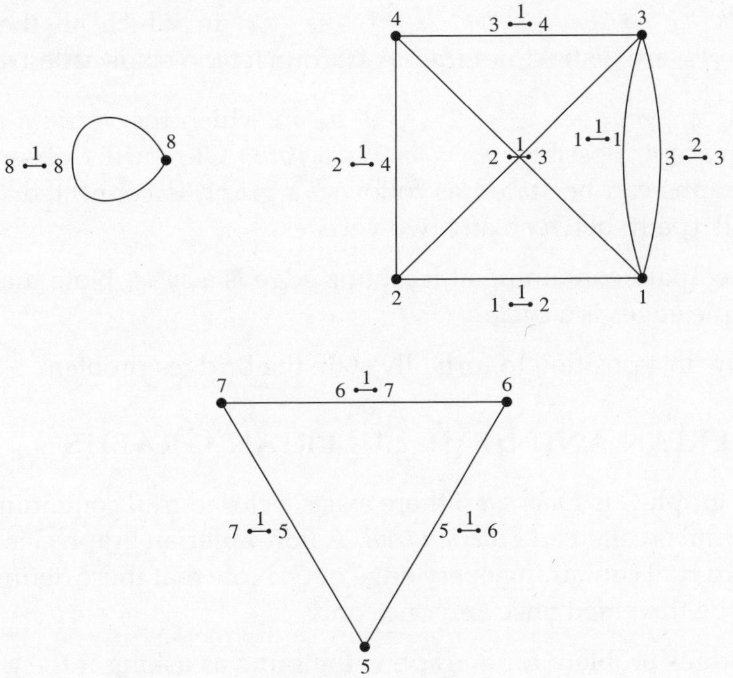

Fig. 3.3.4 Graph 'gr'

Now let us consider the commands that will tell us which vertices are adjacent to a given vertex in the graph "gr". The command `AdjacencyList[gr, 3]` gives the output $\{1,2,4\}$ which tells us that the vertices $1,2,4$ are all adjacent to the vertex 3. Similarly `AdjacencyList[gr, 7]` gives the output $\{5,6\}$.

Investigation 3.12 For the graph in Fig. 3.3.2, find adjacent vertices, adjacent edges, end-vertices, degree of vertices. What is the sum of the degrees of the vertices? What is the size of the edge set? Can the sum of vertex-degrees of a graph be odd? Is there a connection between the sum of vertex-degrees and the size of the edge set? Investigate using `Mathematica` too.

Walk: A walk is a finite sequence of edges of the form $v_0v_1, v_1v_2, \cdots, v_{m-1}v_m$. It is also denoted by $v_0 \to v_1 \to v_2 \to \cdots \to v_m$. Further, any two consecutive edges are adjacent or identical. We call v_0 the *initial vertex* and v_m the *final vertex* of the walk, and speak of a walk from v_0 to v_m. The number of edges in a walk is called its *length*.

In Fig. 3.3.2, $u \to v \to w \to z \to z \to w \to v$ is a walk from u to v of length 6, whereas $u \to v$ is a walk of length 1.

Trail: A *walk* $v_0 \to v_1 \to v_2 \to \cdots \to v_m$, in which all the edges v_0v_1, $v_1v_2, \ldots, v_{m-1}v_m$ are distinct is a *trail*. A trail with $v_0 = v_m$ is called a *closed trail*.

Path: A trail $v_0 \to v_1 \to v_2 \to \cdots \to v_m$ in which the vertices $v_1, v_2, \cdots, v_m$ are distinct (except, possibly, $v_0 = v_m$) is a *path*. Our earlier observation about connected graphs can be stated as follows: a graph is connected if and only if there is a walk (path) between any two vertices.

Cycle: A closed path containing at least one edge is a *cycle*. Note that any loop or pair of multiple edges is a cycle.

We are now in a position to formally state the Bridges problem.

3.4 Eulerian and Semi-Eulerian Graphs

A connected graph G is *Eulerian* if there exists a closed trail containing every edge of G. Such a trail is called an *Eulerian trail*. A non-Eulerian graph G is *semi-Eulerian* if there exists a trail containing every edge of G. Note that these definitions require each edge to be traversed once and once only.

So the Bridges problem for a graph is the same as asking if the given graph is semi-Eulerian and if we consider the modified Bridges problem, where the start and end vertex has to be the same, then the question translates to whether the graph is Eulerian.

As seen at the end of Section 3.2, the results we shall prove are:

Theorem 3.1 (Semi-Eulerian Graphs) A connected graph is semi-Eulerian if and only if it has exactly two vertices of odd degree and the rest of even degree.

Theorem 3.2 (Eulerian Graphs) A connected graph is Eulerian if and only if the degree of each vertex is even.

We shall prove the second Theorem and the first will follow as a corollary. We need the following small result.

Lemma 3.3 If G is a graph in which the degree of each vertex is at least 2, then G contains a cycle.

Proof. If G has any loops or multiple edges, the result is trivial. We can therefore suppose that G is a simple graph. Let v be any vertex of G. We construct a walk $v \to v_1 \to v_2 \to \cdots$ step by step, by choosing v_1 to be any vertex adjacent to v and, for each $i > 1$, choosing v_{i+1} to be any vertex adjacent to v_i except v_{i-1}; the existence of such a vertex is guaranteed by our hypothesis that each vertex has at least two adjacent vertices. Since G has only finitely many vertices, we must eventually choose a vertex that has been chosen before. Let v_k be the first such vertex to be repeated. Let the walk between the two occurrences be $v_k \to v_{k+1} \to \cdots \to v_m = v_k$. Then by the way in which we make the choice of the next vertex, and since v_k is the first repetition we get that the vertices $v_k, v_{k+1}, \ldots, v_{m-1}$ are all distinct as are the edges $v_k v_{k+1}, \ldots, v_{m-1} v_m$. Thus the walk between the two occurrences of v_k will be a closed path and hence we have obtained the required cycle. □

The figure below illustrates a graph with each vertex having degree at least two. In this graph we can see a walk $v_1 \to v_2 \to v_3 \to v_4 \to v_5 = v_2$. Here the required cycle is given by the closed path $v_2 \to v_3 \to v_4 \to v_2$.

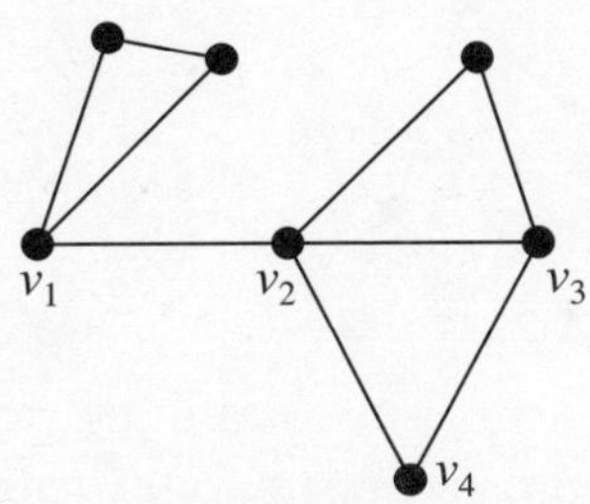

Fig. 3.4.1 Illustration of the Lemma

Proof of Theorem 3.2 (Euler 1736): Let G be a connected graph (with at least two vertices). First we show that if G is Eulerian then each of its vertices must have even degree. Let P be an Eulerian trail of the graph G. Recall that P is a walk starting and ending at the same vertex and passing through each edge of G once and only once. Notice that each vertex v of G can be the start and end point of an Eulerian trail. Thus each vertex must have an edge by which one can leave it and another edge by which we can end there. Thus each vertex will have a degree of at least 2.

Further, whenever P passes through a vertex v, there is a contribution of 2 towards the degree of that vertex as there is an edge via which we reach v and another via which we exit v. In a trail since edges are distinct, every time a vertex occurs in the trail, two such different edges will be counted. Since each edge occurs exactly once in P, each vertex must have even degree.

We prove the converse by induction on the number of edges of G. Suppose that the degree of each vertex is even. Since G is connected, each vertex has degree at least 2 and so, by the Lemma, G contains a cycle C. If C contains every edge of

G, the proof is complete. If not, we remove from G the edges of C to form a new, possibly disconnected, graph H with fewer edges than G and in which each vertex still has even degree. Note that some vertices may now become isolated.

By the induction hypothesis, each component of H has an Eulerian trail. Since each component of H has at least one vertex in common with C, by connectedness, we obtain the required Eulerian trail of G by following the edges of C until a non-isolated vertex of H is reached, tracing the Eulerian trail of the component of H that contains that vertex, and then continuing along the edges of C until we reach a vertex belonging to another component of H, and so on. The whole process terminates when we return to the initial vertex. The three figures below illustrate this process.

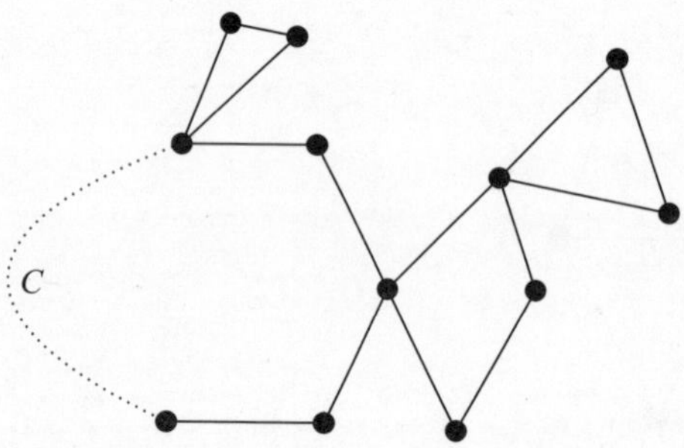

Fig. 3.4.2 Connected Graph G with a cycle C

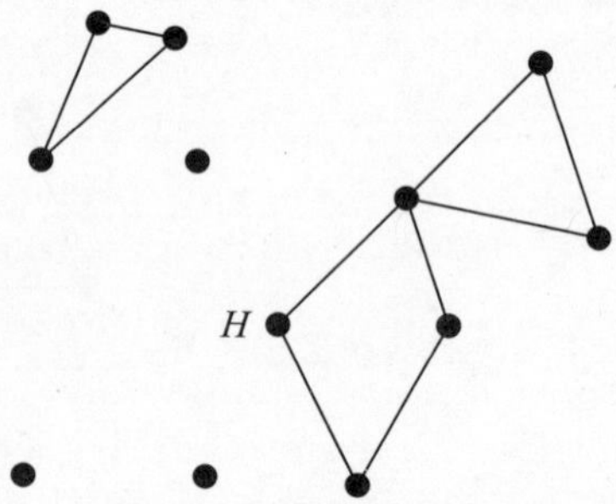

Fig. 3.4.3 H obtained by removing edges of C from G

Figure 3.4.2 shows a connected graph G in which each vertex has an even degree, with a cycle C. Figure 3.4.3 shows the disconnected graph H which we obtain by removing the edges of C from the graph G.

Figure 3.4.4 shows some vertices marked which will help us trace the process of creating the Eulerian trail inductively. We start by finding an Eulerian trail in the top component of H which is $c_1 \to v_2 \to v_3 \to c_1$. Then we continue in the cycle C to build further, $c_1 \to v_2 \to v_3 \to c_1 \to c_4 \to c_5$. Now we have reached the second connected component of H. There we find the Eulerian trail $c_5 \to v_6 \to v_7 \to v_8 \to v_6 \to v_9 \to v_{10} \to c_5$. Now we are back in C and can continue in C to get an Eulerian trail for G given by

$$c_1 \to v_2 \to v_3 \to c_1 \to v_2 \to v_3 \to c_1 \to c_4 \to c_5 \to v_6 \to v_7 \to v_8 \to v_6 \to v_9 \\ \to v_{10} \to c_5 \to c_{11} \to c_{12} \cdots \text{via C} \cdots \to c_1.$$

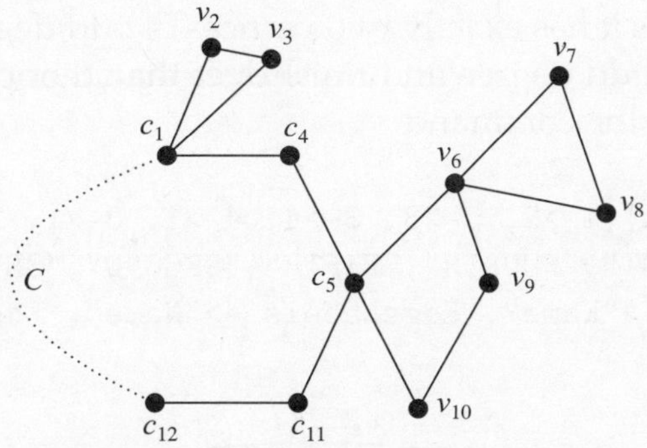

Fig. 3.4.4 G with vertices labelled

We now prove Theorem 3.1 on semi-Eulerian graphs as a corollary to the above theorem.

Let G be a semi-Eulerian graph. So there is a trail P containing each edge of G. Let the initial vertex of P be a and the end vertex be b. Thus P is given by $a \to \cdots \to b$. Now add an extra edge between a and b to get a new graph $\hat{G}$. Clearly $a \to \cdots \to b \to a$ will be an Eulerian trail in $\hat{G}$ where the $a \to \cdots \to b$ is P and the $b \to a$ part is via the new edge added. Thus $\hat{G}$ is an Eulerian graph and so all its vertices have an even degree. But we get G back from $\hat{G}$ by removing that extra edge. So apart from a and b which will now have odd degree in G, all others have even degree.

Now assume that G is a connected graph with exactly two vertices of odd degree, say a and b, and others of even degree. Now add an extra edge say f between a and b to create a new graph $\hat{G}$. Clearly $\hat{G}$ has all vertices of even degree. So we must have an Eulerian trail P in $\hat{G}$. Since P is a closed trail, we can place the vertices occurring in P in sequence, on a circle. Thus we can begin P at any vertex v of $\hat{G}$, and make any edge incident to that vertex v as the last edge taking us back to v. Now assume we have done so by choosing a as our vertex and f as the last edge that brings us back to a. Now we can write P as $a \to \cdots \to b \to a$ where the last $b \to a$ is via the edge $ba = f$. Now remove f, the last edge traversed by P. Then the part left namely $a \to \cdots \to b$ is a trail in G. Note since a and b have odd degree G cannot be Eulerian and so it is semi-Eulerian.

We end this part of the chapter with a discussion on using `Mathematica` to find whether a given graph is Eulerian and, if it is an Eulerian graph, to find an Eulerian trail in it. Consider the graph g in Fig. 3.2.8. To check if this graph is Eulerian we use `EulerianGraphQ[g]`. The output on running this command is "False". Recall that we had found the vertex degrees for g were $\{4, 3, 4, 3\}$. So our Theorem on Eulerian graphs also tells us that this is not an Eulerian graph. However, it is a semi-Eulerian graph as it has exactly two vertices of odd degree. The trail starts in one of the vertices of odd degree and finishes at the other one. The graph below has been produced by the command

```
EdgeTaggedGraph[{1->2, 2->3, 3->4, 4->1, 3->1, 2->4, 1->3} ->
 {"SIXTH","FIRST","FOURTH","FIFTH","SECOND","SEVENTH","THIRD"},
 VertexLabels ->"Name", EdgeLabels ->"Name", EdgeStyle ->{Thick}]
```

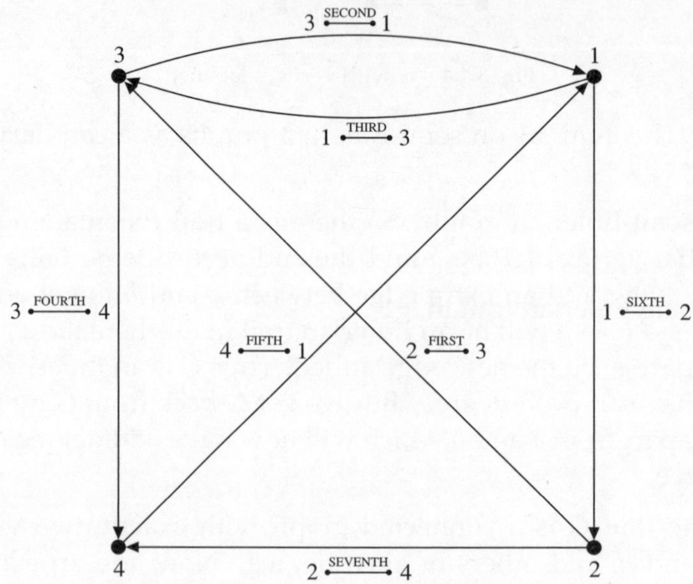

Fig. 3.4.5 Semi-Eulerian Trail in graph 'g'

Note that we are showing the trail by making the edges directed. Further the edges have been labelled by the position they hold in the trail. The trail illustrated above is $2 \to 3$, $3 \to 1$, $1 \to 3$, $3 \to 4$, $4 \to 1$, $1 \to 2$, $2 \to 4$. Also note that there can be several such trails. The proof of the theorem for semi-Eulerian graphs used the theorem for Eulerian graphs. The main step was to convert the two vertices with odd degrees to having an even degree by adding an additional vertex between them. The result is a graph with all degrees even and which is Eulerian. We illustrate this by adding an edge between vertices 2 and 4 of the graph 'g'. The command

```
g1 = EdgeTaggedGraph[{1<->2, 2<->3, 3<->4, 4<->1, 1<->3, 2<->4,
1<->3,  2<->4}, VertexLabels->"Name", EdgeLabels->"Name",
EdgeStyle ->{Thick}]
```

gives the output

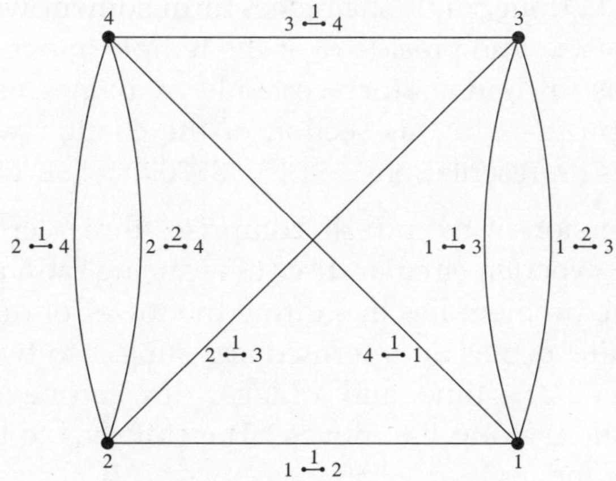

Fig. 3.4.6 Eulerian graph 'g1'

Now `EulerianGraphQ[g1]` returns output 'True'. Further `FindEulerianCycle[g1]` finds the following Eulerian trail in 'g1'.

$$\left\{\left\{1 \xrightarrow{1} 4,\ 4 \xrightarrow{1} 2,\ 2 \xrightarrow{2} 4,\ 4 \xrightarrow{1} 3,\ 3 \xrightarrow{1} 1,\ 1 \xrightarrow{2} 3,\ 3 \xrightarrow{1} 2,\ 2 \xrightarrow{1} 1\right\}\right\}$$

Fig. 3.4.7 Eulerian cycle in 'g1'

Note that `VertexDegree[g1]` gives us the output $\{4, 4, 4, 4\}$ which verifies that we now have a graph in which each vertex has an even degree. While the above trail started with vertex 1, we can rearrange it to start from vertex 2 to get the Eulerian trail $2 \to 4,\ 4 \to 3,\ 3 \to 1,\ 1 \to 3,\ 3 \to 2,\ 2 \to 1,\ 1 \to 4,\ 4 \to 2$. Now, to find a semi-Eulerian trail in the graph 'g', all we need to do is remove the extra edge between vertex 2 and vertex 4. Thus $2 \to 4,\ 4 \to 3,\ 3 \to 1,\ 1 \to 3,\ 3 \to 2,\ 2 \to 1,$ $1 \to 4$ gives us a semi-Eulerian trail for the graph 'g' from vertex 2 to vertex 4, which is different from the one we had earlier.

3.5 The Tower of Hanoi

Those who have heard of the Tower of Hanoi will know that it is a very popular puzzle. It intrigues most people and can be explored in a number of ways. It is also perhaps one of the most celebrated problems in computer science. Discovered by the French mathematician Edouard Lucas in 1883 [53], it is also referred to as the Lucas Tower. There is an interesting legend associated with the puzzle. It goes thus—under the great dome of a temple in Varanasi, there are three gold posts.

One of the posts has 64 gold discs of reducing radii arranged on it, that is, the biggest disc is at the bottom, while the others form a tower with the smallest disc on top. Brahmin priests, who preside over the temple, move the discs according to certain rules. This intriguing story certainly motivates us to do an in depth exploration of this puzzle. In this section of the chapter we shall explore the puzzle, using various representations.

The traditional version of the puzzle comprises three vertical pegs stuck to a wooden base and six wooden circular discs of reducing radii placed on one of the pegs (Fig. 3.5.1). The problem lies in shifting the tower of discs from one peg to another using the third peg as an intermediary, subject to two constraints—only one disc can be moved at a time, and a bigger disc can never be placed on top of a smaller one while making the moves. The shift has to be achieved using a minimum number of moves.

Fig. 3.5.1 A wooden model of the Tower of Hanoi puzzle. Image credit [C3].

If you don't have a physical model of the puzzle, you may try to play the game online, for example at [74].

Investigation 3.13 Try to make the moves with 6 discs and estimate the minimum number of moves that will enable you to transfer all six discs to another peg.

Simplifying the problem

It may be difficult to deal with all six discs at a time. The problem may be simplified by starting with two discs on any peg and counting the minimum number of moves required to shift these discs to any other peg. This may be repeated with 2, 3, 4 and 5 discs. As the minimum number of moves is recorded for various numbers of discs, a pattern is likely to emerge. While exploring the number of moves you may perform the following investigation.

Investigation 3.14 Let n represent the number of discs and $T(n)$ the minimum number of moves required for shifting n discs. Compute $T(n)$ for $n = 2, 3$ and 4. Do you detect any pattern as n increases from 2 to 3? From 3 to 4? Can you express $T(n)$ in terms of $T(n-1)$?

Let us name the three pegs as A, B and C. Note that if two discs are placed on one peg (say peg A) to start with, then the smaller disc can be moved to peg C, the bigger one to peg B and finally the small one can go on top of the bigger one in peg B. Thus a minimum of three moves are required to shift two discs. The moves are illustrated in Fig. 3.5.2. We may represent this as $T(2) = 3$.

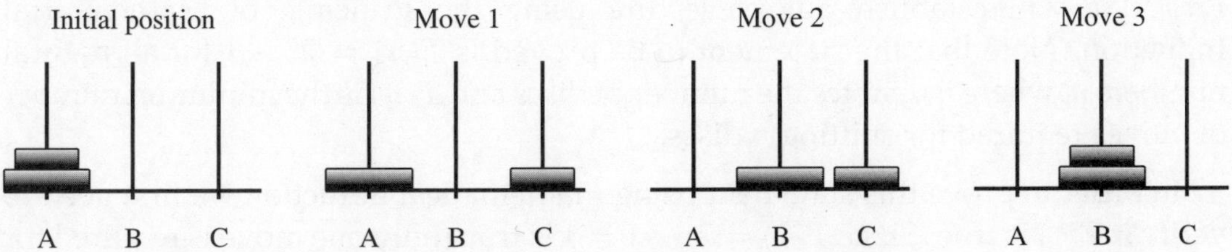

Fig. 3.5.2 Shifting two discs requires 3 moves

Next let us start with three discs on peg A. In order to shift these three discs from peg A to peg B, we need to first shift the two smaller discs (from A to C) in order to release the largest disc. This will require three moves. Once this is done the largest disc can be shifted to B in one move. Finally the other two discs will again take three moves to be placed back on top of the largest disc. Thus the total number of moves for shifting three discs is $3 + 1 + 3 = 7$. The moves are illustrated in Fig. 3.5.3. Try this out with four discs and then five. You will find that the minimum number of moves are $7 + 1 + 7 = 15$ for four discs and $15 + 1 + 15 = 31$ for five discs. If you have noted the pattern you will be able to express $T(n)$ in terms of $T(n-1)$. This is a recursive relationship as it relates the minimum number of moves for n discs with the minimum number of moves for $(n-1)$ discs and can be written as $T(n) = 2T(n-1) + 1$.

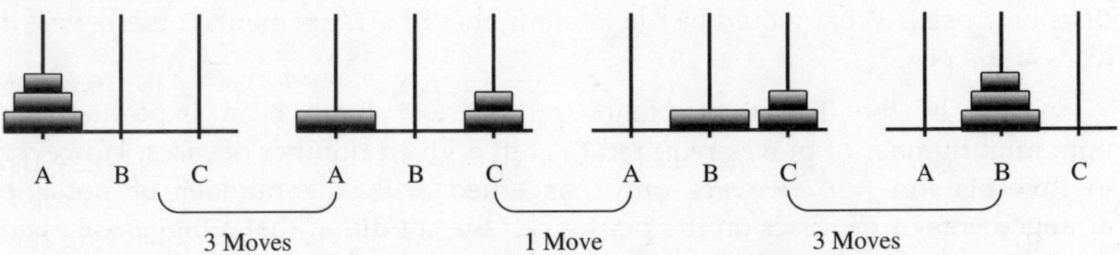

Fig. 3.5.3 Shifting three discs requires 7 moves

Investigation 3.15 Write an explicit rule for $T(n)$ in terms of n. (Hint: Find the nth term of the sequence of number of moves.)

If you have completed Investigation 3.14, you will have observed that $T(n) = 1, 3, 7, 15$ for $n = 1, 2, 3$ and 4 respectively. Using the recursive rule we just found, we can calculate the next few terms as well:

$$T(5) = 2\,T(4) + 1 = 31, \quad T(6) = 2\,T(5) + 1 = 63, \quad T(7) = 2\,T(6) + 1 = 127, \ldots$$

This sequence of numbers motivates the conjecture that $T(n) = 2^n - 1$ for every n.

Proof. We will establish our conjecture using the Principle of Mathematical Induction. Note that the statement to be proved is $T(n) = 2^n - 1$ for all natural numbers n, where n denotes the number of discs and $T(n)$ is the minimum number of moves required for shifting n discs.

In order to prove this statement using mathematical induction we first need to verify if $T(1)$ is true. Since $T(1) = 2^1 - 1 = 1$ is true (only one move is required for shifting one disc), we can proceed with induction. Let us assume $T(k) = 2^k - 1$, for a natural number k. We need to show that $T(k+1) = 2^{k+1} - 1$.

$$T(k+1) = 2T(k) + 1 = 2(2^k - 1) + 1 = 2^{k+1} - 2 + 1 = 2^{k+1} - 1$$

This implies that the statement $T(n) = 2^n - 1$ is true for all natural numbers n. Note that $T(n) = 2T(n-1) + 1$ is the recursive rule whereas $T(n) = 2^n - 1$ is the explicit rule. It is interesting to note that the recursive rule is needed to prove the explicit rule.

Counting problems arising from the ToH puzzle

In this section we shall explore some interesting counting problems, which arise from the puzzle.

Investigation 3.16 In how many ways can two discs be arranged on three pegs? What about three discs? Can we find a rule for the number of arrangements of n discs on 3 pegs? What would be the total number of arrangements if there were d discs and p pegs?

So far, in the Tower of Hanoi puzzle, we have been exploring the minimum number of moves required to shift a given number of discs. However in Investigation 3.16, we are only concerned with the number of possible arrangements of the discs on the pegs given the condition that a bigger disc can never be placed on top of a smaller disc. If there are only two discs, we observe that the bigger disc can be placed in three ways (can be placed on any one of the three pegs) and corresponding to each of these ways, the smaller disc can also be placed in three ways. Thus the number of arrangements should be $3 \times 3 = 9$. Fig. 3.5.4 illustrates the 9 arrangements.

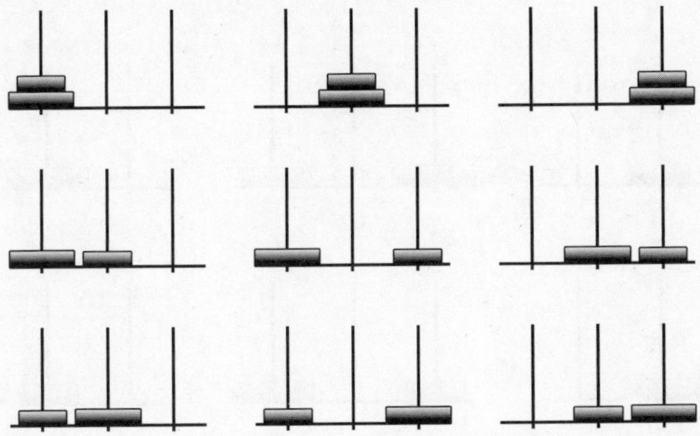

Fig. 3.5.4 A pictorial representation of the arrangements of 2 discs on 3 pegs

3.6 HANOI GRAPHS

By now it must clear to you that pictorial representations are impractical as the number of discs are increased. For example, for 3 discs there are $3^3 = 27$ arrangements. It would be too tedious to represent these pictorially. This motivates the need for a more efficient method of representing the arrangements. In this section we shall try to represent the arrangements using a graph theoretic approach. To identify the nine arrangements or states for two discs (as shown in Fig. 3.5.4), let us number the three pegs with 0, 1 and 2. Each state is assigned two digits based on the pegs on which the discs are positioned, starting with the position of the larger disc. Thus, the state with the larger disc on peg 2 and the smaller one on peg 1, is referred to as 21. If both discs are on peg 0 then this state will be referred to as 00. Figure 3.6.1 shows the nine states along with their respective numbers.

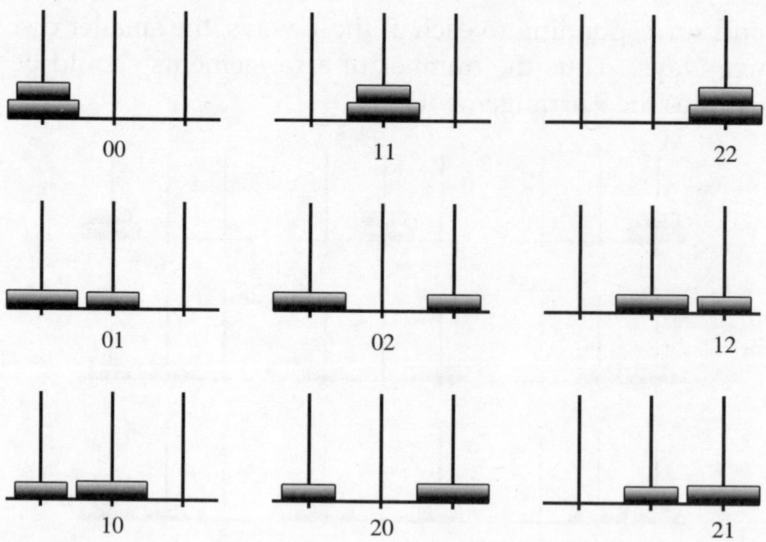

Fig. 3.6.1 The nine states of 2 discs on 3 pegs and their numerical representations

Investigation 3.17 Identify those states, which can be reached from any of the nine states in one move. For example, it is possible to arrive at 10 or 12 from 11 in just one move. (11 implies that both discs are on peg 1. If the smaller disc is shifted to peg 0 we arrive at state 10 and if the smaller disc is shifted to peg 2 we arrive at state 12.)

In this section we shall learn how to represent the states and moves of the ToH puzzle through Hanoi graphs. In a Hanoi graph the states are represented by vertices. Two vertices are joined by an edge if it is possible to shift between those two states using only one move. A Hanoi graph representing d discs and p pegs is denoted by $H(d,p)$. Figure 3.6.2 shows the trivial case of one disc and three pegs, namely, the Hanoi graph $H(1,3)$. Here the vertices represent the three states 0, 1 and 2 respectively. All three vertices are mutually connected by edges leading to a 'triangle'.

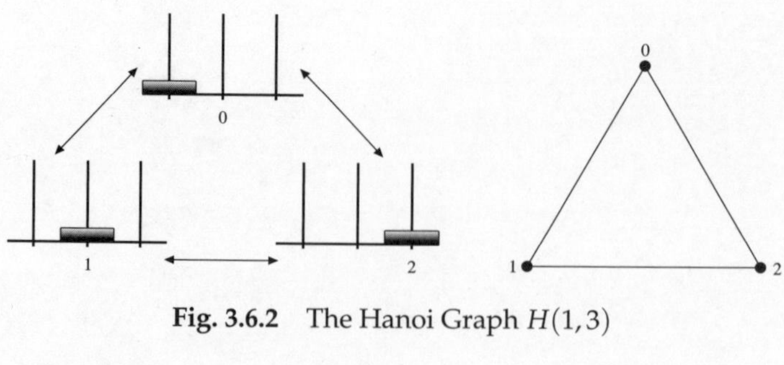

Fig. 3.6.2 The Hanoi Graph $H(1,3)$

Let us now try to construct the Hanoi graph $H(2,3)$. Clearly we need 9 vertices to represent the nine states. Of these 00, 11 and 22 will be referred to as homogeneous states whereas the remaining six states are heterogeneous states. We shall try to maintain a 'triangular arrangement' ensuring that the three homogeneous states 00, 11 and 22 are represented by the three corner vertices of the graph. Note also that from each homogeneous state we can go to only two other states whereas from any non-homogeneous state we can go to three other states. For example, from 00 we can go to 01 and 02 only. However from 12 we can go to 10, 11 or 02. Verify a few other cases for yourself.

In order to draw $H(2,3)$ we will form three triangles, one connecting 00, 01 and 02; a second one connecting 11, 10 and 12; and a third one connecting 22, 20 and 21. These three triangles may be positioned in the top, right and left corners of $H(2,3)$ respectively. Once this is done, the graph comprises three disconnected triangles. Finally we observe that vertex representing 01 may be joined to 21, 02 to 12 and 10 to 20. Why? The completed $H(2,3)$ is shown in Fig. 3.6.3. We find that there are three 'copies' of $H(1,3)$ in $H(2,3)$.

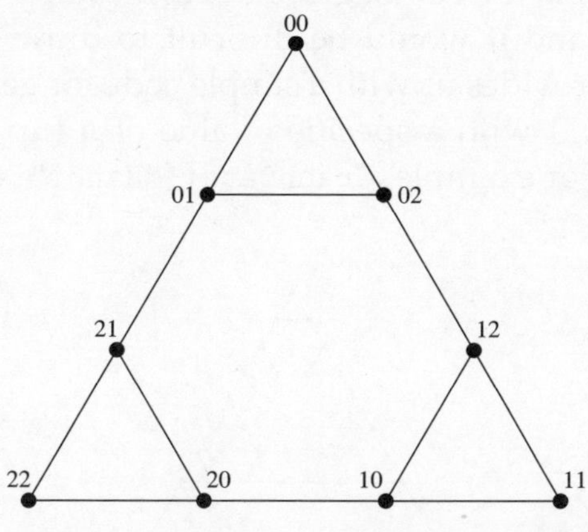

Fig. 3.6.3 Building H(2,3) by connecting three copies of H(1,3)

Hanoi graphs and self-similarity

When you explore fractals in Chapter 5 of this book, you will be excited to see that $H(2,3)$ resembles the Sierpinski triangle. You will be able to identify with the notion of self-similarity when you observe that $H(2,3)$ has three copies of $H(1,3)$. Extending this idea we conjecture that $H(3,3)$ should also have three copies of $H(2,3)$.

Investigation 3.18 Construct $H(3,3)$ using the idea of self-similarity.

From pictorial to numerical representations

Let us revisit the Hanoi graphs $H(2,3)$ and $H(1,3)$. We shall now try to make connections between these graphs and the minimum number of moves required for shifting the discs. Note that the edges create paths along a graph between any two vertices. In particular, let us consider a few paths between the vertices 00 and 11 in $H(2,3)$: 00 $\rightarrow$ 10 $\rightarrow$ 21 $\rightarrow$ 11; 00 $\rightarrow$ 02 $\rightarrow$ 12 $\rightarrow$ 10 $\rightarrow$ 11; 00 $\rightarrow$ 02 $\rightarrow$ 01 $\rightarrow$ 21 $\rightarrow$ 20 $\rightarrow$ 10 $\rightarrow$ 11. Similarly we can find many paths between 00 and 11. However the shortest among these is the path 00 $\rightarrow$ 10 $\rightarrow$ 21 $\rightarrow$ 11 which comprises three edges only and is a part of the boundary of $H(2,3)$. These edges represent the minimum number of moves required for shifting two discs.

Investigation 3.19 Identify the shortest path between two homogeneous vertices in $H(3,3)$ and $H(4,3)$. Note that we can identify and represent paths in $H(4,3)$ even though we haven't constructed it!

Exploring Hanoi Graphs using Mathematica

You may have noticed that the Hanoi graphs become fairly complex as the number of discs are increased and it would be difficult to draw them by hand beyond $H(3,3)$. `Mathematica` provides us with a simple code for generating Hanoi graphs. `GraphData[{"Hanoi",n}]` with a specified value of n (up to 5) will generate the Hanoi graph $H(n,3)$. For example, `GraphData[{"Hanoi",4}]` gives $H(4,3)$ as the output.

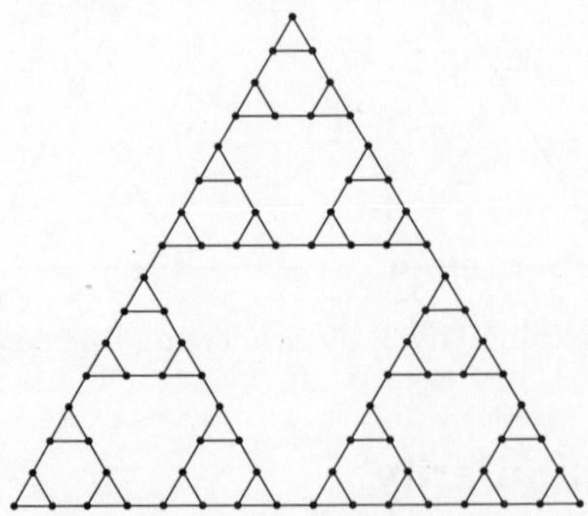

Fig. 3.6.4 `Mathematica` generated H(4,3)

`Table[GraphData[{"Hanoi",n}],n,1,5]` will display the graphs up to $H(5,3)$ simultaneously. The Sierpinski-like structure of the Hanoi graphs are evident from Fig. 3.6.5.

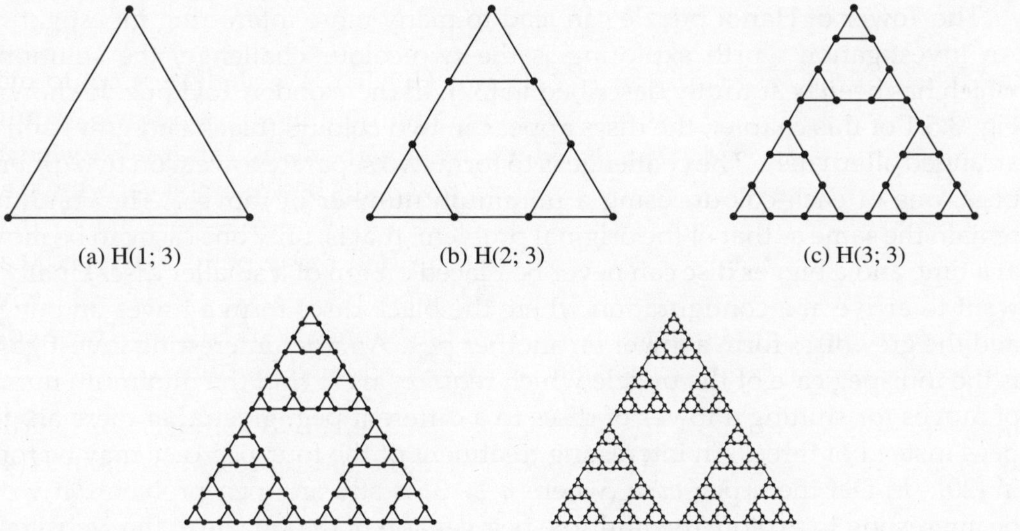

(a) H(1; 3) (b) H(2; 3) (c) H(3; 3)

(d) H(4; 3) (e) H(5; 3)

Fig. 3.6.5

Investigation 3.20 In the previous section you have learnt about Eulerian cycles on graphs. A related concept of Hamiltonian paths is described below under 'Further Explorations'. Can you identify some Eulerian cycles and Hamiltonian paths on the Hanoi graphs?

FURTHER EXPLORATIONS

This chapter began with a puzzle that led to the development of Graph Theory. The paper [19] provides a wonderful analysis of how Euler solved the Bridges problem which was based first on abstracting and generalising the problem. Then there was use of simple logic to analyse a generalised problem to arrive at a solution. The book [4] gives a historical perspective of how graph theory developed and [89] is a good book to learn basic graph theory from. The former traces the connection between KBP and diagram tracing puzzles and includes a translation of Euler's paper [12]. The latter book provides a rich collection of puzzles and other application oriented problems related to graph theory. A closed Hamiltonian trail in a connected graph is a closed trail which goes through every vertex of the graph exactly once. Further exploration of graphs can be done to understand when such Hamiltonian cycles exist. The name associated with such trails came about because of a mathematical game proposed by W R Hamilton and actually marketed in 1859. See [49] for more on Hamilton.

The Tower of Hanoi puzzle can lead to many more interesting investigations. An investigation worth exploring is the two-colour challenge, the solution of which has been beautifully described in [37]. In the wooden ToH puzzle shown in Fig. 3.5.1 of this chapter, the discs appear in two colours (black and grey) and are arranged alternately. The challenge is to form two separate towers on two different pegs, one of each colour, using a minimum number of moves. The conditions remain the same as that of the original problem, that is, only one disc can be moved at a time and a bigger disc can never be placed on top of a smaller disc. Finally we want to arrive at a configuration where the black discs form a tower on one peg and the grey discs form a tower on another peg. Another interesting investigation is the four-peg case of the puzzle which requires us to find the minimum number of moves for shifting a tower of discs to a different peg, given that there are four pegs instead of three. An interesting treatment of the four-peg case may be found in [20]. In fact the n-peg case (where $n > 3$) is still an open problem. It would be interesting to explore the multiple peg version of the puzzle at the website [74] which allows the user to vary the number of pegs and discs and visualise the solution.

4

The Mathematics of Ciphers: Public Key Encryption and RSA

4.1 SECRET MESSAGES

In this chapter we enter into the world of the 'hidden' and learn about the mathematical methods we can use to 'hide'. As children, we have often wanted to send 'secret messages'. The aim was that no one intercepting the secret message should be able to understand the message. Of course, the friend for whom the secret message was written needed to know how to decode the secret message. We shall explore cryptography, the study of the method and techniques to keep communications secure so that only the intended recipients of the communication are able to understand it and no other third party is able to make sense of it.

The word cryptography is made up of two parts: crypto, which comes from the Greek work 'kruptos' meaning hidden and graphy which also derives from the Greek word 'graphia' meaning writing. Thus, cryptography literally means hidden writing. Cipher means secret or disguised way of writing. The main idea here is to write (encrypt) your message in such a way that even if your enemy intercepts the message it is impossible for him to understand the message. But the friend receiving the (encrypted) message should be able to recover the original message from it.

Some terminology that we shall use: *plaintext* will represent the message before encryption; *ciphertext* will be used for the message after encryption. *Encryption* is the process of converting plaintext to ciphertext.

Early historical examples of cryptography that can be found in the literature are the Spartan Skytale and the Caesar Cipher. The Spartan Skytale (also scytale), dating back to about the 5th century BCE, is considered to be one of the first ever devices to be used by the military for encryption. It consists of a wooden rod about which a leather strip is wound. The message is written horizontally across the leather strip that has been wrapped on the rod. After the message is written, the leather strip is unwound from the rod and carries the encrypted message. Figure 4.1.1 shows a picture of a skytale with a leather strip wound on it with a message written.

Fig. 4.1.1 Skytale with leather strip. Image credit [C4].

Figure 4.1.2 shows how the plaintext message 'Help me at once' is encrypted using a skytale.

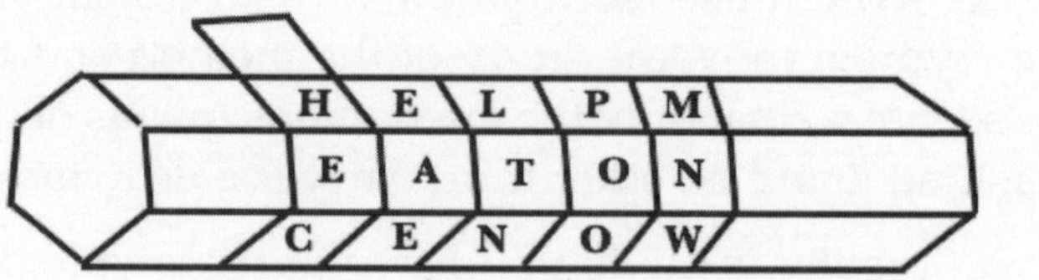

Fig. 4.1.2 Encrypting plaintext using a skytale

Once the leather strip is unwound we get the ciphertext shown in Fig. 4.1.3.

	H	E	C	E	A	E	L	T	N	P	O	O	M	N	W	

Fig. 4.1.3 Ciphertext from a skytale

Only rewinding the strip around another skytale of the same size and shape will reveal the original message.

Caesar's Cipher, also known as Caesar's Shift Cipher or Rolling Code, was supposed to have been used by Julius Caesar in 50 BCE to write secret messages during times of war. The Caesar Cipher works as follows. On the top row we have the plaintext alphabet and on the second row is the ciphertext alphabet indicating the substitution of letters.

Plaintext alphabet:	*A*	*B*	*C*	*D*	*E*	*F*	*G*	*H*	*I*	*J*	*K*	*L*	...
Ciphertext alphabet:	*D*	*E*	*F*	*G*	*H*	*I*	*J*	*K*	*L*	*M*	*N*	*O*	...

So the plaintext 'HELP' will become the ciphertext 'KHOS'. Here the cipher alphabet is created by shifting or rolling the plain alphabet by a certain number of places, 3 to the right in the above example. To recover the plaintext you have to shift the letters in the ciphertext by 3 places to the left.

The two examples illustrated above fall into the two branches into which encryption separates itself historically—transposition ciphers and substitution ciphers. In transposition ciphers, the letters of the plaintext are rearranged by a specified method. Another simple example of a transposition cipher is the 'rail fence' transposition. In this, every other letter of the plaintext message is written in the line below. The method is illustrated via the example shown in Fig. 4.1.4.

Plaintext: HELP ME AT ONCE

Encryption: H L M A O C
E P E T N E

Ciphertext: HLMAOCEPETNE

Fig. 4.1.4 Encryption using Rail Fence Cipher

To recover the plaintext from the ciphertext, the process has to be simply reversed. This is again illustrated in Fig. 4.1.5.

Ciphertext: HLMAOCEPETNE

Decryption:
Step 1: Break into two halves.
HLMAOC
EPETNE
Step 2: Insert the letters from line 2 alternately.
H[E]L[P]M[E]A[T]O[N]C[E]

Plaintext: HELP ME AT ONCE

Fig. 4.1.5 Decryption of a Rail Fence Cipher

The second type of cipher is the substitution cipher. Here we substitute for each letter of the alphabet, another letter of the alphabet via a specified method. The Caesar Cipher is an example of this method. Note that in the transposition cipher, each letter of the alphabet remains itself, but its position in the plaintext gets changed to form the ciphertext. However, in a substitution cipher, the position of each letter between the plaintext and ciphertext stays the same but each 'letter' has a new identity!

4.2 ENCRYPTION SCHEMA

It is time to step back from the actual methods used and come up with a schematic for the main ingredients that are required to create an encryption system. But the path to this is through a more detailed breakdown of the steps. Let us consider the Caesar Cipher. The main method here is a 'rolling or shift' substitution of the letters. In our example, the shift was by 3 places to the right. We can think of this as the *encryption key* and similarly, the *decryption key* would be to shift 3 places to the left.

So the main ingredients are–plaintext, an algorithm or method to encrypt the plaintext, an encryption key which controls the actual encryption, ciphertext, a decryption key that controls the actual decryption and an 'inverse' algorithm by which the decryption is done.

Thus we have two significant parts in an Encryption process (Fig. 4.2.1).

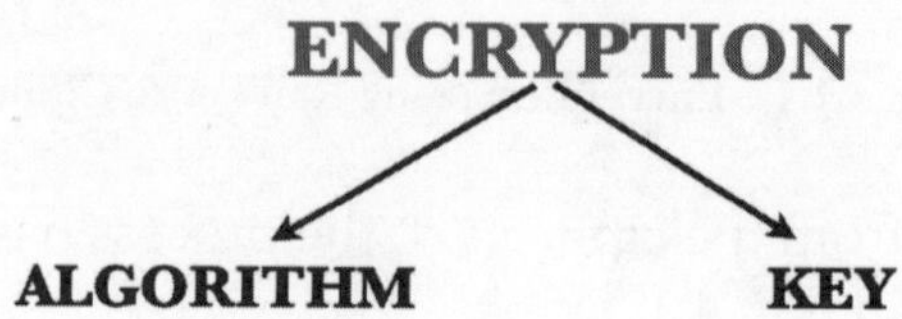

Fig. 4.2.1 Encryption

Let us explore what these two parts are for the examples we have seen in the previous section.The algorithm is basically the encryption method.

Spartan Skytale: The algorithm here is to wrap a strip of leather around the skytale and write the message horizontally.

Caesar's Shift Cipher: The algorithm here is to shift the plaintext alphabet by a certain number of places to obtain the ciphertext.

Rail Fence Cipher: The algorithm here is to write some 'specified' letters of the plaintext in a line below and then to write the new message as line 1 followed by line 2. The key specifies the exact details of the particular encryption.

Spartan Skytale: The key here is the specific skytale used.

Caesar's Shift Cipher: The key here is the number of places the plaintext alphabet is shifted. In our example the key was 3.

Rail Fence Cipher: The key here specifies which letters are written in the line below. In our example the key is that every alternate letter is written on the line below.

The schema for encryption and decryption is summarised in Fig. 4.2.2.

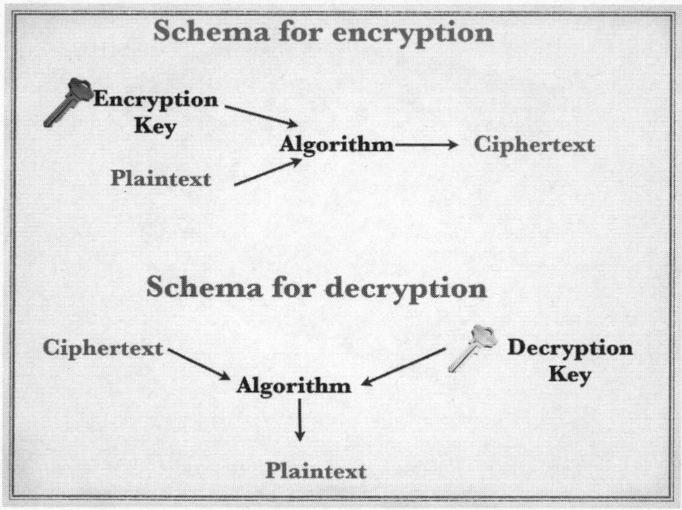

Fig. 4.2.2 Encryption and decryption schematic

In the above examples, the encryption key and the decryption key are very closely connected. Indeed, so closely that, knowing one would allow us to find the other easily. Such encryption systems are called *symmetric key* encryption systems. Such systems lead to a conundrum as far as encryption is concerned. If the intended recipient of your encrypted message has to decrypt the message, she needs the decryption key. So how do you pass on the decryption key to her in secret? The decryption key also needs to be encrypted! And we end up with a chicken and egg situation unless there was a prior secret meeting to decide on the encryption and decryption key. How do you resolve such a problem?

Let us investigate via the puzzle below and maybe we can develop some insight.

Investigation 4.1 You are in a country where the only way to send items securely is to put them in a box with an old fashioned padlock and send the box by post. Below are some more details.

(i) You are also allowed to put as many locks as you like.

(ii) Any item posted without being inside a padlocked box will be stolen.

(iii) There is no other way to send items in that country.

(iv) The law of the country is that no one is allowed to break such a box or the lock, and all people living in the country follow the law.

(v) See Fig. 4.2.3 for an image of such a box.

Fig. 4.2.3 Padlocked box. Image credit [C5].

(vi) Uttara and Neelambari live in two different parts of the country.

(vii) Uttara has to send a small bag of precious stones to Neelambari who lives in a far off part of the country.

(viii) Uttara has put the small bag in a padlocked box and posted it to her friend Neelambari.

(ix) Neelambari receives the box but it is locked and she does not have the key.

Can you think of a way to help Uttara and Neelambari out so that Neelambari is able to recover the bag of precious stones, within the law of the land? Are you able to connect this with the conundrum of passing on the decryption key?

In 1976, in an invited paper, Diffie and Hellman [9] proposed a new system called 'public key encryption' which used asymmetric keys. The idea was that one could develop encryption systems in which the encryption key and decryption key are connected (they reverse each other's effect) but knowing one would not allow you to find the other in real time, even using a powerful computer. The term 'public' is to indicate that even if one of the keys was in the public domain, the other private or secret key cannot be found using the public key. They, however, did not give a concrete example of any public key encryption in the paper. But in the same paper, they proposed a method in which the symmetric keys of a conventional encryption system could be 'exchanged' securely over insecure channels. This procedure is now know in the literature as the Diffie–Hellman Key Exchange. Figure 4.2.4 shows a snapshot of their original paper.

644 IEEE TRANSACTIONS ON INFORMATION THEORY, VOL. IT-22, NO. 6, NOVEMBER 1976

New Directions in Cryptography

Invited Paper

WHITFIELD DIFFIE AND MARTIN E. HELLMAN, MEMBER, IEEE

***Abstract*—Two kinds of contemporary developments in cryptography are examined. Widening applications of teleprocessing have given rise to a need for new types of cryptographic systems, which minimize the need for secure key distribution channels and supply the equivalent of a written signature. This paper suggests ways to solve these currently open problems. It also discusses how the theories of communication and computation are beginning to provide the tools to solve cryptographic problems of long standing.**

Fig. 4.2.4 Diffie and Hellman's 1976 paper

Imagine a phone directory with the names of all the people who wish to receive messages. Against each of their names will be an encryption key denoted by e. Anyone wishing to send Aslam a message will use the e in front of Aslam's name to encrypt the message. Only Aslam knows the secret decryption key d that will decrypt the message. This is an illustration where the encryption key e for sending the message to Aslam is public knowledge. It can be used by anyone who wishes to send Aslam a secret message. The encrypted message can even be posted on a public bulletin board for Aslam. But only Aslam who has the secret decryption key d can use it to decipher the secret message. This concept of public key encryption caught the imagination of cryptographers and within two years, in 1978, Rivest, Shamir and Adleman [60] had created a practical method for public key encryption which is now known as *RSA*. The phrase 'RSA' comes from the first letters of their names. They exploited a simple fact that we have all experienced in mathematics at school, namely it is easier to multiply two numbers rather than divide or factorise a given number which is a product of two numbers.

Investigation 4.2 The Mathematica input code `Times[a, b, c, ..., z]` gives the output '$abc \cdots z$', the product of the numbers $a, b, c, \ldots, z$. The command `FactorInteger[n]` gives output of the form '$\{\{p_1, k_1\}, \{p_2, k_2\}, \ldots, \{p_t, k_t\}\}$'. This output has to be interpreted as giving the prime factorisation of the integer n: $n = p_1^{k_1} p_2^{k_2} \cdots p_t^{k_t}$.

(a) Complete the following manually and then solve using `Mathematica`.

(i) Let $n = pq$ where $p = 43$ and $q = 53$. Find n.

(ii) Let $n = 3869$. Factorise n.

(b) Try to find the prime factorisation of 235764327 by hand or a calculator (and it will be great if you did succeed!) and then use `Mathematica`.

We are sure that the answer to the first question will take you less than a minute whereas the answer to the second would take you a lot longer even with a calculator. In the next section we will find out more about the mathematics used for RSA.

But before that, a slight aside. It was found out much later that James Ellis of the British Secret Service had come up with the idea of public key encryption in the 1960s. This was much before Diffie and Hellman but alas for national security reasons it was never announced to the public. For more on this, read the article [28] in Wired magazine.

4.3 Modular Arithmetic and the Euler Phi Function

In this section we will learn to do modular arithmetic or what is also known as clockwork arithmetic. This sort of arithmetic is required to understand RSA. Let us start with a simple example. Suppose you know that the first of March is a Sunday and you are asked to find the dates on which Sundays will occur in the month. You would be able to answer this easily—Sundays in March will be on $1, 8, 15, 22$ and 29. What was the method used? You simply added 7 to 1 to get 8. Then 7 added to 8 gave you 15 and so on. Another way to look at this is to find the dates (numbers) in the month that leave the remainder 1 on division by 7. Notice that

$$\begin{aligned} 1 &= 7 \times 0 + 1 \\ 8 &= 7 \times 1 + 1 \\ 15 &= 7 \times 2 + 1 \end{aligned}$$

$$\begin{aligned} 22 &= 7 \times 3 + 1 \\ 29 &= 7 \times 4 + 1 \end{aligned}$$

We write this as $29 \equiv 1 \bmod 7$ or $1 \equiv 1 \bmod 7$ and read it as 29 is congruent to 1 modulo 7. What this means is that 29 leaves the remainder 1 on division by 7.

Let n be a natural number and let a and b be integers. We say that $a \equiv b \bmod n$ if n divides $a - b$. Below are some properties that modular arithmetic follows and which we will use.

(a) Reflexivity: $a \equiv a \bmod n$.

(b) Symmetry: $a \equiv b \bmod n$ implies $b \equiv a \bmod n$.

(c) Transitivity: $a \equiv b \bmod n$, $b \equiv c \bmod n$ implies $a \equiv c \bmod n$.

(d) Let $a \equiv b \bmod n$ and $c \equiv d \bmod n$ and let $k \in \mathbb{N}$. Then

  (i) Additivity: $a + c \equiv b + d \bmod n$.

  (ii) Multiplicativity: $ac \equiv bd \bmod n$.

  (iii) Power: $a^k \equiv b^k \bmod n$.

The above properties are not difficult to prove. We prove (d)(ii). The rest are left for the reader to prove. Let $a \equiv b \bmod n$ and $c \equiv d \bmod n$. So $a - b = qn$ and $c - d = tn$ for some $q, t \in \mathbb{Z}$. Note that

$$\begin{aligned} ac - bd &= ac - bc + bc - bd \\ &= c(a-b) + b(c-d) \\ &= cqn + btn \\ &= (cq + bt)n \end{aligned}$$

Since $h = cq + bt \in \mathbb{Z}$ and $ac - bd = hn$ we see that n divides $ac - bd$ and so $ac \equiv bd \bmod n$.

We learnt at school that if we divide a natural number a by a natural number n we get a unique quotient q and a unique remainder r. The method we used to find the quotient and remainder was the long division method. Euclid's Division Algorithm (EDA) states this in more generality. Let $n \in \mathbb{N}$. For any $a \in \mathbb{Z}$ there exist unique integers q and r such that $a = qn + r$ and $0 \leq r \leq n - 1$. We can write this as $a \equiv r \bmod n$. Note that r is the unique remainder we get on dividing a by n. We also write $r = a \bmod n$. The properties of additivity, multiplicativity and power give us the following. (You can find out more about Euclid at [45].)

Let $a, c \in \mathbb{Z}$ and let $n, k \in \mathbb{N}$. Then modulo n we have,

(i) Additivity: $(a + c) \bmod n \equiv (a \bmod n) + (c \bmod n)$.

(ii) Multiplicativity: $(ac) \bmod n \equiv (a \bmod n)(c \bmod n)$.

(iii) Power: $a^k \bmod n \equiv (a \bmod n)^k$.

Also note that the only remainders we can get on dividing by n are the numbers $0, 1, \ldots, n-1$. This is what makes modular arithmetic powerful. When we want to do arithmetic modulo n, then we only need to work with the remainders and not the entire collection of integers. Let us illustrate this further. Suppose we want to find $(3^7 + (-5)^{51}) \bmod 7$. Then using additivity, multiplicativity, power and other properties of modular arithmetic we get the following modulo 7.

$$\begin{aligned}
(3^7 + (-5)^{51}) \bmod 7 &\equiv (3^7 \bmod 7) + ((-5)^{51} \bmod 7) \\
&\equiv \{(3^2 \bmod 7)^3 (3 \bmod 7)\} + \{2^{51} \bmod 7\} \\
&\quad \text{since } -5 \bmod 7 = 2 \\
&\equiv \{(2 \bmod 7)^3 (3 \bmod 7)\} + \{(2^3)^{17} \bmod 7\} \\
&\quad \text{since } 3^2 \bmod 7 = 2 \\
&\equiv (3 \bmod 7) + (1 \bmod 7) \\
&\quad \text{since } 2^3 \bmod 7 = 1 \\
&\equiv 4
\end{aligned}$$

Thus, 4 will be the remainder left on dividing $(3^7 + (-5)^{51})$ by 7 or $(3^7 + (-5)^{51}) \bmod 7 = 4$.

Investigation 4.3 The `Mathematica` input code `Mod[a,n]` gives the output 'r' where $r = a \bmod n$. Note that in `Mathematica`, `3^7` is the input code for 3^7. Use `Mathematica` to find $(3^7 + (-5)^{51}) \bmod 7$.

Investigation 4.4 Do the following using modular arithmetic and also using `Mathematica`.

(a) Show that $(1051 + 2733) \bmod 13 = ((1051 \bmod 13) + (2733 \bmod 13)) \bmod 13$.

(b) Show that $(1051 \times 2733) \bmod 13 = ((1051 \bmod 13) \times (2733 \bmod 13)) \bmod 13$.

(c) Verify that $(1051)^5 \bmod 13 = ((1051 \bmod 13)^5) \bmod 13$.

`Mathematica` also has the input code `PowerMod[a, b, m]` which gives as output $a^b \bmod m$. This can also be used.

The next mathematical tool we study is called *Euler's phi function*, named after the 18th century mathematician Leonhard Euler. More about Euler's life and work can be found at [46].

We define Euler's phi function as follows. Let n be a positive integer. Then $\phi : \mathbb{N} \to \mathbb{N}$ is given by

$\phi(n) =$ The number of positive integers less than or equal to n and coprime to n.

Let us examine some examples and properties of the ϕ function. It is clear that $\phi(1) = 1$. For any $n > 1$, since $\gcd(n, n) = n > 1$, to find $\phi(n)$, we only need to find the number of positive integers that are less than n and relatively prime to n. As 1 is the only number less than and coprime to 2 we get $\phi(2) = 1$.

Now, $1, 2$ are both less than and coprime to 3. Thus $\phi(3) = 2$. Similarly, only $1, 3$ are less than 4 and coprime to 4. Therefore $\phi(4) = 2$. All natural numbers less than 5 are coprime to 5 so $\phi(5) = 5 - 1 = 4$. Indeed for any prime p, we have $\phi(p) = p - 1$, as all the natural numbers less than p will be coprime to p.

Let p be a prime. How do we find $\phi(p^n)$? Note that the only numbers that are less than p^n and not coprime to p^n are those which are multiples of p. That is, the numbers $p, 2p, 3p, \ldots, p^{n-1}p$. But these are p^{n-1} in all. So the number of positive integers less than p^n and coprime to p^n will be $p^n - p^{n-1}$. Thus $\phi(p^n) = p^n - p^{n-1} = p^{n-1}(p - 1)$, for p prime.

A very important property of the ϕ function is that it is multiplicative. This means, if m, n are natural numbers and $\gcd(m, n) = 1$ then $\phi(mn) = \phi(m)\phi(n)$. The proof can be found in [6].

This property allows us to find $\phi(n)$ if we know the prime decomposition of n.

Let $n = p_1^{k_1} p_2^{k_2} \cdots p_t^{k_t}$ be the prime decomposition for n. Since the ϕ function is multiplicative we get

$$\begin{aligned} \phi(n) &= \phi(p_1^{k_1})\phi(p_2^{k_2}) \cdots \phi(p_t^{k_t}) \\ &= (p_1^{k_1} - p_1^{k_1-1})(p_2^{k_2} - p_2^{k_2-1}) \cdots (p_t^{k_t} - p_1^{k_t-1}) \end{aligned}$$

Below is an illustrative example.

$$\begin{aligned}\phi(1400) &= \phi(2^3 \times 5^2 \times 7)\\ &= (2^3 - 2^2) \times (5^2 - 5) \times 6\\ &= 4 \times 20 \times 6\\ &= 480\end{aligned}$$

A special case of the multiplicative property of $\phi(n)$ gives us the following.

If p and q are distinct primes then,

$$\phi(pq) = (p-1) \times (q-1) \tag{4.1}$$

Investigation 4.5 The `Mathematica` input code `EulerPhi[n]` gives as output the value of $\phi(n)$. Use `Mathematica` to find $\phi(53 \times 37 \times 24)$.

Investigation 4.6 Do the following manually and also using `Mathematica`.

(a) Find the prime factorisation of $n = 596232$.

(b) Find $\phi(n)$.

(c) Verify that if $n = p_1^{k_1} p_2^{k_2} \cdots p_t^{k_t}$ is the prime factorisation found in (a) then $\phi(n) = (p_1^{k_1} - p_1^{k_1-1})(p_2^{k_2} - p_2^{k_2-1}) \cdots (p_t^{k_t} - p_1^{k_t-1})$.

4.4 Some Important Results

We are now in a position to state three important results—Fermat's Little Theorem, its generalisation Euler's Theorem and the Modular Inverse Theorem.

Theorem 4.1 (Modular Inverse Theorem) Let n and a be positive integers such that $\gcd(a, n) = 1$. Then there exists a positive integer a' such that $aa' \equiv 1 \bmod n$.

Proof: Since $\gcd(a, n) = 1$, there exist integers m and a' such that $mn + aa' = 1$. But this is the same as saying $aa' \equiv 1 \bmod n$. Thus, we have proved the Modular Inverse Theorem. □

Without a loss of generality, we can assume that $1 \le a' \le n$ as $r \equiv a' \bmod n$ will also satisfy $ar \equiv 1 \bmod n$. When we find a' such that $1 \le a' \le n$ and $aa' \equiv 1 \bmod n$ then we call a' the 'inverse of a modulo n'.

The Modular Inverse Theorem tells us that given an a which is coprime to n, its inverse modulo n exists. How do we find such an a'? Below, we provide an illustrative method using an easy example.

The aim is to find the inverse of 64 modulo 11. Thus we need to find an a' such that $64 \times a' \equiv 1 \bmod 11$. We do so by the repeated use Euclid's Division Algorithm. The method is illustrated below. In Step 1, the larger number 64, is divided by the smaller number 11 to obtain the quotient 5 and remainder 9.

Step 1: $64 = (5 \times 11) + 9$

In the next step we divide 11 (the divisor in Step 1) by 9 (the remainder in Step 1).

Step 2: $11 = (1 \times 9) + 2$

In the next step we divide 9 (the divisor in Step 2) by 2 (the remainder in Step 2).

Step 3: $9 = (4 \times 2) + 1$

The division process stops as the remainder in Step 3 is 1, which is $\gcd(64, 11)$. This, in fact, is the long division method taught in school to find the gcd of two numbers.

The aim via the next few steps is to write the gcd of 64 and 11 as a linear combination $1 = (64 \times a') + (11 \times m)$ for some integers a' and m.

Step 4: $1 = 9 - (4 \times 2)$ by rewriting Step 3.

Step 5: $1 = 9 - \{4 \times \{11 - (1 \times 9)\}\}$, using Step 2 to substitute for 2.

Step 6: So $1 = 9 - (4 \times 11) + (4 \times 9) = (-4 \times 11) + (5 \times 9)$.

Step 7: $1 = (-4 \times 11) + \{5 \times \{64 - (5 \times 11)\}\}$, using Step 1 to substitute for 9.

Step 8: So $1 = (-4 \times 11) + (5 \times 64) - (25 \times 11) = (64 \times 5) + (11 \times -29)$.

Step 9: Thus $64 \times 5 = 1 \bmod 11$ or the inverse of 64 modulo 11 is 5.

Investigation 4.7 The `Mathematica` command `QuotientRemainder[a,n]` gives as output '$\{q, r\}$' where q is the quotient and r is the remainder given by $a = qn + r$ in Euclid's Division Algorithm. Find the inverse of 75 modulo 7 manually and using the above `Mathematica` command.

`Mathematica` also has several commands to find 'inverse modulo n' directly. One is the `PowerMod[a, b, m]` command we saw earlier. The input code `PowerMod[a, -1, m]` will give the inverse of a modulo m, if $\gcd(a, m) = 1$, otherwise it returns a message 'a is not invertible modulo m'. Another command that can be used is `ModularInverse[k, n]` to get the inverse of k modulo n.

Fermat's Little Theorem, also known as FLT is not the famous FLT which stands for Fermat's Last Theorem. Fermat lived in 17th century France and was in fact a lawyer by profession. He is well-known for the famous FLT which states that for $n > 2$, the equation $x^n + y^n = z^n$ has no solutions in positive integers. (For more about Fermat see [48].) In our daily lives though, it is Fermat's Little Theorem that is more useful.

Theorem 4.2 (Fermat's Little Theorem) Let p be a prime and let a be a positive integer such that p does not divide a. Then $a^{p-1} \equiv 1 \bmod p$.

Another equivalent version of the above FLT is the following. Let p be a prime and let a be any positive integer then $a^p \equiv a \bmod p$.

Theorem 4.3 (Euler's Theorem) Let n and a be positive integers such that $\gcd(a, n) = 1$. Then $a^{\phi(n)} \equiv 1 \bmod n$.

Note that if n is prime then Euler's Theorem gives us Fermat's Little Theorem.

This next discussion is only for those who are familiar with group theory. Euler's Theorem (and hence Fermat's Little Theorem) are proved using group theory. Other readers may skip till the illustrative examples. A brief definition of a group is given in the last section of this chapter.

Proof of Euler's Theorem: Lagrange's Theorem states that for a subgroup H of a finite group G, we have that $|H|$ divides $|G|$. In particular, if $a \in G$, then $o(a) = |\langle a \rangle|$ divides $|G|$ or equivalently $a^{|G|} = 1$ where 1 denotes the identity of the group. (For more on Lagrange see [52].)

If n is a positive integer, then the set $U(n) = \{r \in \mathbb{N} \mid 1 \leq r \leq n, \gcd(r, n) = 1\}$, forms a group under multiplication modulo n. Further, $|U(n)| = \phi(n)$. Now consider any $a \in \mathbb{N}$ such that $\gcd(a, n) = 1$. Let $r = a \bmod n$. Then $r \in U(n)$ and as seen above, $r^{\phi(n)} = 1$ in $U(n)$. In other words, we have $r^{\phi(n)} \equiv 1 \bmod n$. By the power property of modular arithmetic we therefore get $a^{\phi(n)} \equiv r^{\phi(n)} \equiv 1 \bmod n$.

Investigation 4.8 Answer the following using `Mathematica`.

(a) Is 9091 prime?

(b) If 9091 is prime, verify FLT for $p = 9091$ and $a = 11$.

(c) Factorise $n = 100001$.

(d) For $n = 100001$, find $\phi(n)$ and verify the multiplicative property of $\phi(n)$.

(e) For $a = 53$ and $n = 100001$, verify Euler's Theorem.

Investigation 4.9 The `Mathematica` input code `GCD[a, b, ...]` gives as output the value of 'gcd$(a, b, \ldots)$'. Let $n = 560027$ and let $a = 1874$. Answer the following using `Mathematica`.

(a) Verify that $\gcd(560027, 1874) = 1$.

(b) Let $a = 1874$. Find its inverse modulo $n = 560027$.

(c) Factorise $n = 560027$.

(d) For $n = 560027$, find $\phi(n)$ and verify the multiplicative property of $\phi(n)$.

(e) For $a = 1874$ and $n = 560027$, verify Euler's Theorem.

4.5 THE RSA ALGORITHM

As mentioned in an earlier section, in 1978, Rivest, Shamir and Adleman proposed a scheme for a public key encryption. This was the asymmetric key cipher which is known as RSA. The asymmetry in the keys is produced using the simple fact that it is easier to multiply than to factorise. This was seen in Investigation 4.2 earlier. So how does RSA work? The mathematics involved is fairly simple. In RSA, the encryption key $\mathbf{e}$ is public as is the number n. The decryption key $\mathbf{d}$ and the distinct primes p, q are secret. The connection is that the factorisation of n is $n = pq$ and $\mathbf{d}$ is the inverse of $\mathbf{e}$ modulo $\phi(n)$.

A person, say Angus, wishes to have secret messages sent to himself. He chooses two distinct primes p and q. These primes are secret and known only to Angus. (In RSA implementation in real life, these primes are over 200 digits. So large primes and the quest for them is essential for encryption in current times.) Angus then multiplies p and q to find $n = pq$. The multiplication process is easy and is implementable using computers. Since Angus knows the primes p and q, by using Eq. (4.1) he can find $\phi(n) = (p-1)(q-1)$. Now he chooses a number $\mathbf{e}$ such that $\gcd(\mathbf{e}, \phi(n)) = 1$. This $\mathbf{e}$ will serve as the public encryption key. Now, by the Modular Inverse Theorem, $\mathbf{e}$ has an inverse modulo $\phi(n)$, namely $\mathbf{d}$. Since Angus knows $\phi(n)$, he can find $\mathbf{d}$ and this will be the secret decryption key known only to Angus.

Note that the information that Angus will put out in public is n and $\mathbf{e}$. The information known only to Angus and kept secret is $p, q, \phi(n)$ and $\mathbf{d}$. While the precise method of RSA will be explained later, let us examine whether the private or secret information can be found from the public information.

To find $\mathbf{d}$ as the inverse of $\mathbf{e}$ modulo $\phi(n)$, we need to know the value of $\phi(n)$. We show that knowing $\phi(n)$ is mathematically equivalent to factorising n as $n = pq$. Thus cracking RSA, is equivalent to being able to factorise numbers, specially those which are known to be a product of two large primes. Factorisation, even with the current level of computational power we have, is difficult for numbers that are a product of two large primes. We will see more about this towards the end of this section in the write-up on RSA Factoring Challenge.

As seen earlier, if we know the factorisation of n, that is if we we know p, q such that $n = pq$, then we know $\phi(n)$, as $\phi(n) = (p-1)(q-1)$. Now, conversely suppose that we know the value of $\phi(n)$. We show that we can find the p, q such that $n = pq$. We have two unknowns, namely p and q, but we do have some information connecting them. The first is that $pq = n$. We also know the value of $\phi(n)$ and we know that $\phi(n) = (p-1)(q-1)$ even though we do not know p or q. Thus we get $\phi(n) = pq - (p+q) + 1 = n - (p+q) + 1$. Thus we get the following two equations in the two unknowns p and q.

$$pq = n \tag{4.2}$$

$$p + q = n - \phi(n) + 1 \tag{4.3}$$

So p and q are the roots of the quadratic equation given by

$$x^2 + (\phi(n) - n - 1)x + n = 0$$

Hence we get

$$p = \frac{-b + \sqrt{b^2 - 4c}}{2} \text{ and } q = \frac{-b - \sqrt{b^2 - 4c}}{2}$$

where $b = \phi(n) - n - 1$ and $c = n$.

Investigation 4.10 Let $n = 2279$ be a product of two distinct primes p, q. Let $\mathbf{e} = 73$. Then do the following.

(a) Use `Mathematica` to find $\phi(n)$.

(b) Use the method above to find p and q.

(c) Use `Mathematica` to factorise n and verify that $n = pq$.

(d) Use `Mathematica` to find **d**.

Now we describe the steps involved in RSA encryption and decryption. Let us assume that Angus has to receive a secret message from Bulku. The process for key generation is described below.

Key Generation

(i) Angus chooses two distinct primes p and q which are kept secret and announces publicly the number n which is the product of p and q.

(ii) Angus then calculates $\phi(n) = (p-1)(q-1)$ which is also kept secret. He then chooses a number **e** such that $\gcd(\mathbf{e}, \phi(n)) = 1$. Usually **e** is also chosen to be a prime not dividing $\phi(n)$. This **e** is the encryption key and is made public by Angus.

(iii) Angus calculates the secret decryption key **d** as the inverse of **e** modulo $\phi(n)$.

Below are the steps involved in the conversion of secret message that Bulku wants to send to Angus. Recall that we use the term plaintext to refer to the secret message before it is encrypted. After encryption, the secret message is called ciphertext. The RSA algorithm referred to below is described later.

Plaintext to Ciphertext and Back

(i) The plaintext is translated to a plaintext number, say M. This is done by assigning numbers to every letter in the alphabet. For example, the ASCII code assigns numbers to all keyboard characters. (See `https://theasciicode.com.ar/`.)

(ii) The plaintext number is encrypted to ciphertext number, say C, using the RSA algorithm and the **public encryption key**. This is done by Bulku using the public encryption key **e** displayed publicly by Angus.

(iii) The ciphertext number (and the ciphertext) are posted publicly by Bulku, say on a public bulletin board, as a message for Angus.

(iv) The ciphertext number is decrypted to the plaintext number using the RSA algorithm and the **secret decryption key**. This is done by Angus using the secret decryption key **d** that only he knows.

(v) The decrypted plaintext number is translated back to the plaintext.

The RSA algorithm is described below. This is being done in the context of Angus wishing to receive a secret message from Bulku. Recall that Angus has displayed in public two numbers, namely, n and **e**. For the RSA algorithm to work, the plaintext number M obtained in Step 1 previously, has to satisfy $M < n$. If the plaintext is long, it can be broken into several plaintext messages, such that their corresponding plaintext numbers, say M_i, for $i = i, \ldots, t$, are each numbers less than n. These can be sent sequentially by the process described above. So we shall assume that $M < n$.

RSA Algorithm

(i) For **encrypting** the plaintext number to the ciphertext number, the plaintext number M is raised to the power **e** and reduced modulo n. This gives us the ciphertext number C. Mathematically, $M^{\mathbf{e}} \equiv C \bmod n$. This part is done by Bulku.

(ii) This part is done by Angus. For **decrypting** the ciphertext number to the plaintext number, the ciphertext number C is raised to the power **d** and reduced modulo n. This gives us the plaintext number M. Mathematically, $C^{\mathbf{d}} \equiv M \bmod n$. Since we had assumed that $M < n$, we get that $C^{\mathbf{d}} \bmod n = M$. Thus, the plaintext number is recovered from the ciphertext number.

Note that if we had not insisted on $M < n$, then the above process would have given us a number R which would be the unique remainder when $C^{\mathbf{d}}$ is divided by n. And unless $M < n$, we would have $M \neq R$ and so we would not have recovered our plaintext number and hence our original plaintext message.

Next we see why the RSA algorithm works. As seen in Step 1 above $C \equiv M^{\mathbf{e}} \bmod n$. Thus using modular arithmetic we get,

$$\begin{aligned} C^{\mathbf{d}} &\equiv (M^{\mathbf{e}})^{\mathbf{d}} \bmod n \\ &= M^{\mathbf{ed}} \bmod n \\ &= M^{1+t\phi(n)} \bmod n, \text{ since } \mathbf{ed} \equiv 1 \bmod \phi(n) \end{aligned}$$

Thus we have

$$C^{\mathbf{d}} \equiv M(M^{\phi(n)})^{t} \bmod n \tag{4.4}$$

We have to consider two cases to complete the proof, but before that an observation that t in the equation above will be positive as $\mathbf{ed} > 1$. The first case is when $\gcd(M,n) = 1$. Then by Euler's Theorem, we have $M^{\phi(n)} \equiv 1 \bmod n$. Thus from Eq. (4.4) we get

$$C^{\mathbf{d}} \equiv M \bmod n$$

The second case is when $\gcd(M,n) \neq 1$. Now p, q are distinct primes and $n = pq$. Further $M < n$. Thus, if $\gcd(M,n) \neq 1$, then we must have $\gcd(M,n) = p$ or $\gcd(M,n) = q$. Without loss of generality let us assume that $\gcd(M,n) = q$. Then p does not divide M and so by Fermat's Little Theorem we have $M^{(p-1)} \equiv 1 \bmod p$ and so $(M^{(p-1)})^{(q-1)} \equiv 1 \bmod p$. Since $\phi(n) = (p-1)(q-1)$ we get $M^{\phi(n)} \equiv 1 \bmod p$. From Eq. (4.4), we know that n divides $C^{\mathbf{d}} - M(M^{\phi(n)})^{t}$. Since p divides n,

$$\begin{aligned} C^{\mathbf{d}} &\equiv M(M^{\phi(n)})^{t} \bmod p \\ &\equiv M \bmod p \qquad \text{since } M^{\phi(n)} \equiv 1 \bmod p \end{aligned}$$

Thus,

$$C^{\mathbf{d}} \equiv M \bmod p \tag{4.5}$$

Also since $\gcd(M,n) = q$, we have that $q|M$ and so $M \equiv 0 \bmod q$. Hence again from Eq. (4.4) and since q divides n we get

$$\begin{aligned} C^{\mathbf{d}} &\equiv M\left(M^{\phi(n)}\right)^t \bmod q \\ &\equiv 0 \bmod q && \text{since } M \equiv 0 \bmod q \\ &\equiv M \bmod q && \text{since } M \equiv 0 \bmod q \end{aligned}$$

Thus

$$C^{\mathbf{d}} \equiv M \bmod q \tag{4.6}$$

Equations (4.5) and (4.6) tell us that both p and q divide $C^{\mathbf{d}} - M$. Since p and q are distinct primes, $n = pq$ will also divide $C^{\mathbf{d}} - M$. Consequently, we have $C^{\mathbf{d}} \equiv M \bmod n$ as required.

We now illustrate the RSA encryption and decryption process with an example with some small primes. This is only for the purpose of giving a hands-on feel for the process. Clearly, the choices are such that the encryption can be easily cracked. In real life to implement RSA we need to choose very large primes and the implementations are done using computers. The encryption is real-time robust as there aren't good enough computer programs for real-time factorisation.

Public Directory of RSA

Names	**n**	**e**
Angus	143	17
Bulku	100761443	10007

From the above, since we have chosen small primes it is easy to see that for Angus, $n = 143 = 13 \times 11$ and so $\phi(n) = 12 \times 10 = 120$. Thus $\mathbf{e} = 17$ is certainly coprime to $\phi(n) = 120$. Recall that $\phi(n) = 120$ as well as the decryption key $\mathbf{d} = 113$ is secret information known only to Angus. Recall that `ModularInverse[17, 120]` can be used to find **d**.

Now Bulku wishes to send the plaintext message Save Me. The ASCII printable characters take values from 032 to 126. So we will work with the premise that every letter, character in the keyboard is represented by a 3-digit number. Using the ASCII code this plaintext is translated to the plaintext number $M = 83971181013277101$.

Now recall that the plaintext number M must be less than $n = 143$. So we need to break our message M into blocks. Each block uses as many digits, from left to right, till we get the largest number less than $n = 143$. So here we get 7 blocks, namely $M_1 = 83, M_2 = 97, M_3 = 118, M_4 = 101, M_5 = 32, M_6 = 77, M_7 = 101$. Now we apply the RSA encryption process on the M_i.

Bulku needs to calculate $C_i = (M_i)^{\mathbf{17}} \bmod 143$ for $i = 1, \ldots 7$. She can do it with `Mathematica` as follows:

```
e = 17;
Mi = {83, 97, 118, 101, 32, 77, 101};
Ci = PowerMod[Mi, e, 143]
```

This gives

$$C_1 = 96, \quad C_2 = 15, \quad C_3 = 79, \quad C_4 = 95, \quad C_5 = 54, \quad C_6 = 77, \quad C_7 = 95.$$

Thus the ciphertext number is $C = 96157995547795$. To convert the ciphertext number to ciphertext we need to ensure that the blocks are of size below 256 as the extended ASCII Table has conversion for values from '00 to 256'. We make blocks using as many digits from left to right so that we get the largest number less than 256. We see the corresponding blocks will be 96 157 99 55 47 79 5. Using the ASCII Table in `https://theasciicode.com.ar/` corresponding ciphertext is `Øc/O. Note that 5 stands for the ASCII control 'Enquiry' and so is not represented in the ciphertext. It is important to post the ciphertext number and not just the ciphertext as the ciphertext may miss some control characters which are not written.

Bulku posts the ciphertext number $C = 96157995547795$ on a public bulletin board as a message for Angus. Now Angus has to decrypt C and so he makes blocks. Each block uses as many digits, from left to right, till we get the largest number less than $n = 143$. Thus he will get back the blocks $C_1 = 96, C_2 = 15, \ldots C_7 = 95$. Now, he finds $M_i = (C_i)^{\mathbf{113}} \bmod 143$ for $i = 1, \ldots, 7$, by using `PowerMod[Ci, 113, 143]` to calculate the M_i. The values are as shown below.

$$M_1 = 83, \quad M_2 = 97, \quad M_3 = 118, \quad M_4 = 101, \quad M_5 = 32, \quad M_6 = 77, \quad M_7 = 101$$

Thus the plaintext number we get after decryption is 83971181013277101, which is the M we started with. The corresponding plaintext is found again by making blocks from left to right using as many digits till we get the largest number less than 255. Here the blocks will be 83 97 118 101 32 77 101 which stands for Save Me.

Note that to be able to use RSA, we need to have access to primes. The `Mathematica` command `Prime[n]` gives as output the nth prime p_n. So this can be used to find primes.

Let us work through an example where you wish to enter your name and data in the RSA Public Directory. Below are all the steps that YOU will need to undertake.

Step 1: Choosing the primes p and q (kept secret) and calculating $n = pq$ (made public).

```
p = Prime[50]
q = Prime[100]
n = p*q
```

This gives $p = 229$, $q = 541$ and $n = pq = 123889$.

Step 2: Calculating $\phi(n)$ (kept secret) and choosing a prime **e** (made public) which is coprime to $\phi(n)$. The number **e** is the public encryption key.

```
phi = EulerPhi[n]
FactorInteger[phi]
```

The first command gives $\phi(n) = 123120$. The output of the second command is `{{2,4},{3,4},{5,1},{19,1}}` and this gives the factorisation,

$$\phi(n) = 2^4 \times 3^4 \times 5 \times 19$$

So YOU need to choose **e** to be a prime different from 2, 3, 5 and 19. A safe choice is to choose the twentieth prime:

```
e = Prime[20]
```

This sets $\mathbf{e} = 71$.

Step 3: Entry in the RSA Public Directory

Names	**n**	**e**
YOU	123889	71

Step 4: Calculating the decryption key **d** (kept secret).

```
d = ModularInverse[71, 123120]
```

We get $\mathbf{d} = 102311$

Investigation 4.11 Cheeni has sent YOU a secret message. She has posted the following Ciphertext number $C = 10103911264451987$ on the RSA Public Bulletin Board. Find the corresponding ciphertext and her secret message. Translate her secret message (plaintext) to the plaintext number M and verify using the RSA encryption process that C is the ciphertext number obtained.

Investigation 4.12 Recall that Bulku has $n = 100761443$ and $\mathbf{e} = 1007$ in the RSA Public Directory. You want to send Bulku a secret message. The secret message is 'Run away'. Do the following.

(a) Translate the plaintext into a plaintext number using the ASCII Table.

(b) Encrypt the plaintext number to a ciphertext number.

(c) Translate the ciphertext number to ciphertext using the ASCII Table.

(d) Find Bulku's secret decryption key. Use it to decrypt the ciphertext number. Verify that the plaintext number and plaintext are recovered.

We end this section with a note on the RSA Factoring Challenge.

RSA Factoring Challenge

RSA Security LLC (see `https://www.rsa.com/`), named after its founders Rivest, Shamir and Adleman was founded in 1982. It is currently part of Dell Technologies. In March 1991, the RSA company announced *The RSA Factoring Challenge*. The challenge offered a cash prize for solving the factorisation of a semiprime (product of two primes). The first semiprime announced was RSA-100, which was a 100-digit semiprime. The prize for cracking it is was 1000 US dollars. As per the Wikipedia page [88], the last semiprime to be cracked was RSA-250 and this was on 28 February 2020. RSA-250 has 250 digits in decimal representation and 829 in binary representation. The highest announced prize was 200,000 US dollars for anyone cracking RSA-2048 which is a semiprime with 617 digits in decimal representation and 2048 in binary representation.

The aim of the challenge was to improve research into computational number theory and to improve integer factorisation algorithms. The challenge officially ended in 2007. It is believed that if quantum computing is achieved successfully then integer factorisation will not be a challenge computationally and thus RSA will no longer be secure.

4.6 Diffie–Hellman Key Exchange

We end this chapter with a short section on the Diffie–Hellman Key Exchange. Recall Investigation 4.1. In the investigation, Uttara had to send a valuable item to Neelambari. It had to be sent only in a locked box to ensure its safety. This created a problem for Neelambari as she is not allowed to break the lock to recover the valuable item. She needs the 'key' to open the lock but since the 'key' is valuable it also needs to be sent in a locked box to Neelambari.

The problem of using a symmetric key encryption system is akin to the above problem. Uttara encrypts the secret message and sends it to Neelambari. Encryption is like the lock in the above puzzle. To decipher the message Neelambari needs the decryption key which is also secret and known only to Uttara. To send the secret decryption key, Uttara has to encrypt it! Also in a symmetric key encryption system, the encryption key also has to be kept secret as the encryption and decryption keys are so closely linked that from the knowledge of one the other can be found.

The Diffie–Hellman Key Exchange protocol is a way in which two persons, say Angus and Bulku, the sender and the receiver, can mathematically arrive at a 'common key' which can be used for the encryption and decryption process. This is done by exploiting the fact that finding 'discrete logarithms' is computationally difficult.

We know that in real numbers we can solve the following problem easily. Given a positive real number g and another positive real number y, we can solve for x in the equation $g^x = y$ by taking logarithms. We know g and y, and so we solve for x and get

$$x = \log y / \log g$$

The discrete version of this problem is the following. Let p be a prime and consider $U_p = \{1, 2, \ldots, p-1\}$. We can define a special way to multiply two numbers in U_p using modular arithmetic. For $r, s \in U_p$, we define the product $r * s = rs \bmod p$. It can be proved that the following hold with respect to $*$.

(a) Closure: For all $r, s \in U_p$ we have $r * s \in U_p$.

(b) Associativity: For all $r, s, t \in U_p$ we have $(r * s) * t = r * (s * t)$.

(c) Existence of Identity: There exists $1 \in U_p$ such that $r * 1 = r = 1 * r$ for all $r \in U_p$.

(d) Existence of Inverses: For each $r \in U_p$ there exists $s \in U_p$ such that $r * s = 1 = s * r$.

Any non-empty set X with an operation $*$ between its elements that satisfies the four properties above is called a **group**.

The above set U_p is a group and is called the group of integers under multiplication modulo p where p is prime. Notice that, $r * s = s * r$ in U_p and so we say that U_p is an abelian group. If for $r, s \in U_p$ we have $r * s = 1$, then we say that s is the inverse of r and denote it as r^{-1}.

For a positive integer m and any $g \in U_p$, we define $g^m = g * g * \cdots * g$ (m times). We define $g^0 = 1$ and when $n = -m$ for a positive integer m, we define $g^n = (g^{-1})^m$. It can be proved that there exists $g \in U_p$ such that every element of U_p can be written uniquely as g^m for some $1 \leq m \leq p-1$. We therefore say that U_p is a **cyclic group** with generator g and denote it by $U_p = \langle g \rangle$. Note that U_p has order $p-1$, that is, it has $p-1$ elements. This is denoted as $|U_p| = p-1$. Let us illustrate with an example.

For example, let $p = 7$. Then $U_7 = \{1, 2, \ldots, 6\}$. Since in U_7, under multiplication modulo 7, we have

$$
\begin{aligned}
3^1 &= 3 \\
3^2 &= 2 \\
3^3 &= 6 \\
3^4 &= 4 \\
3^5 &= 5 \\
3^6 &= 1
\end{aligned}
$$

Thus $U_7 = \langle 3 \rangle$. Now if $y \in U_p = \langle g \rangle$, and p is a large prime, it is almost impossible to solve the equation $g^x = y$, for x. The **discrete logarithm** of y to the base g is x if $g^x = y$. Solving for x is akin to being able to take logarithms in this finite discrete system. This is called the **Discrete Log** problem.

Diffie and Hellman exploited the fact that it would be practically impossible to compute g^{ab} from knowing g^a and g^b. To do so would be like solving the discrete log problem.

Assume that Angus and Bulku want a symmetric key K to use together but the channel of communication they have is not secure. The following steps are taken in the Diffie–Hellman Key Exchange protocol.

- Angus and Bulku publicly announce a large prime p and a generator g for U_p.

- Angus chooses a large number a (kept secret, $a < p$) and announces A publicly, where $A = g^a \bmod p$.
- Bulku chooses a large number b (kept secret, $b < p$) and announces B publicly, where $B = g^b \bmod p$.

Now both Angus and Bulku can generate the same secret symmetric key K as follows

- Angus computes $B^a = g^{ab} \bmod p$.
- Bulku computes $A^b = g^{ab} \bmod p$.
- Thus Angus and Bulku both have a 'common key', namely, $K = g^{ab} \bmod p$.

To find K, an eavesdropper will have to solve the discrete log of A to the base g or the discrete log of B to the base g, which is computationally difficult.

Further Explorations

There are several books that can provide more information about both the history of cryptography as well as other mathematical methods of cryptography. One such book which any informed lay person can read for a good understanding is [66]. The author, Simon Singh, writes books related to science and mathematics. For a more mathematical insight into the encryption methods [6] is recommended. The earlier sections in this chapter showed how RSA could be implemented or the steps worked out using `Mathematica`. A similar approach can be tried for the Diffie–Hellman Key Exchange protocol.

We showed in the section on RSA that the encryption was robust as long as it took a computer a long time to factor a number which is product of two large primes. This was also the basis for the RSA factoring challenge (see [88]). Similarly the Diffie–Hellman Key Exchange Protocol is robust as long as the discrete log problem cannot be tackled by computers in reasonable time. What does 'reasonable time' mean mathematically? How can one mathematically analyse whether a given encryption system is robust or not? Another aspect is how easy or hard is it to implement such protocols on computers. Further, what happens to encryption systems like RSA if faster computers, say computers based on quantum computing (see [87]) become a reality? To answer or explore these one requires some knowledge of topics that lie within the interface of computer science, mathematics and physics. An interview [7] discusses a few consequences of quantum computing.

5

Exploring Fractals

'Clouds are not spheres, mountains are not cones, coastlines are not circles and bark is not smooth, nor does lightning travel in a straight line.' This observation by Benoit Mandelbrot aptly introduces a 'fractal'. The term was first introduced by him in his book *Fractals: Form, Chance and Dimension* [30] published in 1977. It is based on the latin word *fractus* which means broken or fractured. The reader may wish to learn more about Benoit Mandelbrot at [54].

Fractals are found everywhere in nature. Let us see the picture of the fern leaf as shown in Fig. 5.0.1. The main branch comprises smaller leaves which are almost replicas of the main leaf. Further the smaller leaves comprise still smaller ones which resemble the bigger leaf. This example of a pattern, which repeats itself, where each smaller leaf is a scaled down copy of the larger leaves, illustrates the notion of self-similarity. Similarly, if we look closely at a cauliflower, we find that smaller pieces of it resemble the original cauliflower. Here again, we observe a self-similar structure.

Fig. 5.0.1 A fern leaf has leaflets which are similar to the whole leaf. Similarly, each portion of a cauliflower appears as a miniature copy of the whole object. Image credit [C6, C7, C8].

There are many other examples of fractals in nature. However, in this chapter we will focus on some geometrical constructions, which lead to fractal patterns. To put it simply, fractals are never-ending patterns that repeat themselves at different scales. Simple geometrical constructions when replicated lead to complex patterns, which lead to interesting explorations. The purpose of this chapter is to explore some of these well-known fractals. We shall use `MS Excel` and `Mathematica` to observe their mathematical properties.

5.1 THE SIERPINSKI TRIANGLE

The first fractal we shall explore is the Sierpinski triangle (also known as the Sierpinski Gasket or Sierpinski Sieve),which was described by the Polish mathematician Waclaw Sierpinski in 1915. More details about Sierpinski may be found at [56]. The Sierpinski triangle is one of the first fractals ever defined.

To create the Sierpinski triangle we need to cut out an equilateral triangle from a coloured sheet of paper. We mark this as stage 0 of the fractal construction process. To obtain stage 1, we join the midpoints of the three sides to obtain four smaller triangles and remove the triangle at the centre. The resulting piece consists of three smaller equilateral triangles with a triangular 'hole' in the centre (Fig. 5.1.1). To obtain stage 2 we repeat the process of joining the midpoints of the three smaller equilateral triangles, each time removing the centre triangles. This process may be continued indefinitely. Figure 5.1.1 shows stages 0, 1, 2 and 3 of the Sierpinski triangle.

Fig. 5.1.1 Stages 0, 1, 2 and 3 of the Sierpinski triangle

The `Mathematica` command `SierpinskiMesh[n]` may be used to generate the nth stage of the Sierpinski triangle. The command `Table[SierpinskiMesh[n], {n, 0, 3}]` will generate stages 0 to 3 of the fractal.

Investigation 5.1 How many black triangles will there be at the 4th and 5th stages? Can you find a rule for the number of black triangles at the nth stage? If the area of the triangle at stage 0 is 1 square unit, what is the shaded area at stages 1, 2 and 3? At the nth stage?

Stage	0	1	2	3	$\cdots$	n
Number of black triangles						
Shaded (black) area						

The Idea of Self-similarity

Note that stage 1 of the Sierpinski triangle comprises three identical smaller copies of stage 0 where each copy is a smaller equilateral triangle. Similarly stage 2 has three copies of stage 1 and nine copies of stage 0. Can you identify these?

Investigation 5.2 How many copies of stages 0, 1, 2 and 3 do you expect to find in stage 4 of the Sierpinski triangle? In general how many copies of the previous stages will you find in stage n?

Recursive and Explicit Formulae

You may have noted in Investigation 5.1 that the number of shaded triangles leads to the geometric sequence $1, 3, 3^2, 3^3, \ldots$. To find the next term of the sequence we need to multiply the previous term by 3. Thus, if S_n and S_{n-1} represent the number of shaded triangles at stages n and $n-1$ respectively, then $S_n = 3S_{n-1}$. This is the recursive rule or formula for the number of shaded triangles. Hence $S_1 = 3S_0$, $S_2 = 3S_1$, $S_3 = 3S_2$ and so on.

Thus, the explicit rule or formula for finding the number of shaded triangles at the nth stage is 3^n. While filling the table in Investigation 5.1, you may have noted that the exponent of 3 coincides with the number of the stage.

Let us now consider the shaded areas in the different stages of the Sierpinski triangle construction. Observe that in stage 1 the original triangle is divided into 4 congruent smaller triangles and the central one is removed. Thus, if the area of the triangle at stage 0 is 1 square unit then the area at stage 1 is $3/4$ square units. We may conclude that the shaded area in the different stages forms the geometric sequence $1, 3/4, (3/4)^2, (3/4)^3, \ldots$. Here the multiplying factor is $3/4$.

Investigation 5.3 Find the recursive and explicit rules for the shaded area at the nth stage. Let A_n denote the area at the nth stage.

Investigation 5.4 What happens to the number of shaded triangles and the shaded area as n approaches infinity?

A Spreadsheet Exploration

We shall now explore the Sierpinski triangle construction process numerically by generating the data in `Excel`. In Column A of the spreadsheet we shall represent the stage numbers n and in columns B and C we shall represent the number of shaded triangles and the shaded areas respectively. To begin with, enter 0 in cell `A2` (of column A) and enter the formula `=A2+1` in cell `A3`. Drag the cell `A3` to generate the stage numbers till 20. In column B, we enter 1 in cell `B2` and the formula `=B2*3` in cell `B3`. Double click the cursor in the corner of the cell `B3` to generate the sequence of number of shaded triangles. Similarly, enter 1 in cell `C2` and the formula `=C2*(3/4)` in cell `C3`. Double clicking the cell `C3` will generate the sequence of shaded areas.

Investigation 5.5 Generate the data in `Excel` and observe what happens to the column values as the stages increase. Use the Graphing feature of `Excel` to generate the graphs for the number of triangles and the shaded area.

The Sierpinski Carpet

In this section we shall try to extend the idea of the Sierpinski triangle construction to a square. Let us begin with a square sheet of unit length at stage 0. To construct stage 1 we shall trisect each side of the square piece and join the points of trisection of opposite sides. This creates nine smaller squares. We remove the centre square and retain the 8 smaller shaded squares. It appears as though the square sheet has a square 'hole' in the centre. The same process is repeated on the eight smaller shaded squares to obtain stage 2. At stage 2 there are $8 \times 8 = 64$ smaller shaded squares. The process can be continued indefinitely and the figure thus generated is known as the Sierpinski carpet (Fig. 5.1.2).

Fig. 5.1.2 Stages 0, 1 and 2 of the Sierpinski carpet

The `Mathematica` command `MengerMesh[n]` may be used to generate the nth stage of the Sierpinski carpet.

You may have guessed that the number of shaded squares leads to the geometric sequence $1, 8, 8^2, 8^3, \ldots$.

Investigation 5.6 Find the recursive and explicit rules for the number of shaded squares as well as shaded area at the nth stage. Let S_n and A_n denote the number of shaded squares and the shaded area respectively at the nth stage.

Investigation 5.7 What happens to the number of shaded squares and the shaded area as n approaches infinity?

Investigation 5.8 Generate an `Excel` Sheet to explore the Sierpinski carpet.

5.2 THE KOCH SNOWFLAKE

In this section, we shall explore yet another fractal known as the Koch snowflake. It is named after the Swedish mathematician Helge Von Koch who first described the fractal in a paper in 1904. The reader may find more details about Koch at [51].

Stage 0 of this fractal comprises an equilateral triangle. To obtain stage 1 we trisect each side of the equilateral triangle, remove the middle segment and raise smaller equilateral triangles, of side length, one-third of the original triangle, on the middle segment. Thus each side of the original equilateral triangle is replaced by four smaller line segments to obtain stage 1. This process is repeated on every line segment of stage 1 to obtain stage 2, and so on. Figure 5.2.1 shows stages 0, 1 and 2 of the Koch snowflake.

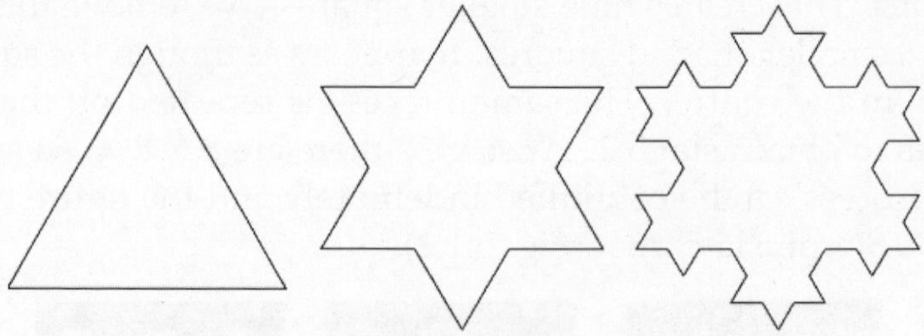

Fig. 5.2.1 Stages 0, 1 and 2 of the Koch snowflake

The Mathematica command `Graphics[KochCurve[n]]` creates the nth stage of this process applied to a single side of the original triangle. The following figure shows the output of `Graphics[KochCurve[3]]` and `Graphics[KochCurve[4]]`:

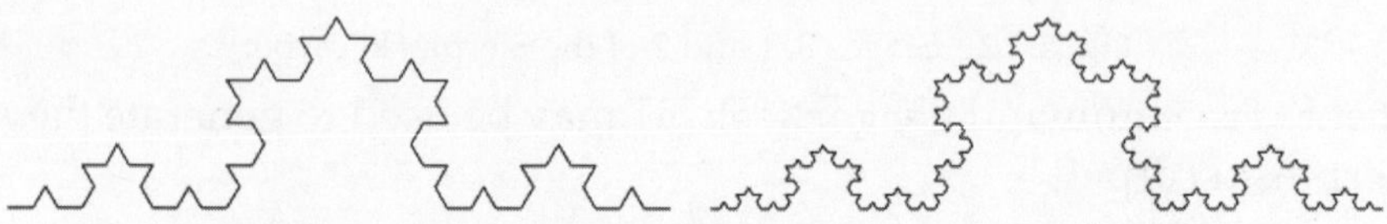

Fig. 5.2.2 Stages 3 and 4 of one side of the Koch snowflake

The full snowflake may be generated by taking two more copies of the nth stage of one side and rotating them about their endpoints. The following code demonstrates this and generates stages 3 and 4 of the Koch snowflake:

```
Table[Graphics[GeometricTransformation[KochCurve[i],
  {RotationTransform[Pi, {1/2, 0}],
   RotationTransform[-Pi/3, {1, 0}],
   RotationTransform[Pi/3, {0, 0}]}]], {i, 3, 4}]
```

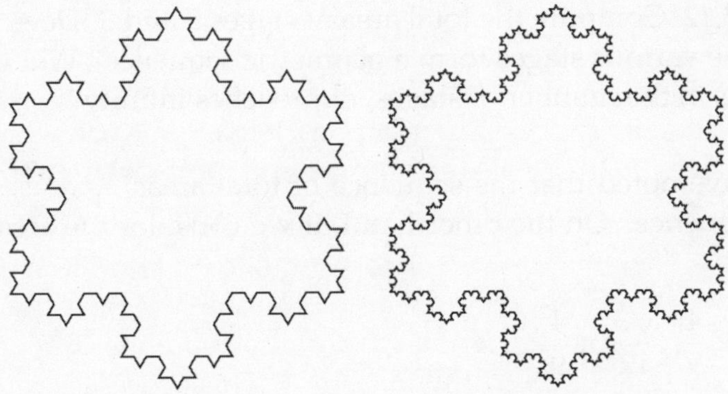

Fig. 5.2.3 Stages 3 and 4 of the Koch Snowflake

Investigation 5.9 Count the number of line segments in stages 0, 1 and 2. Predict the number of line segments at stages 3 and 4. Which constant multiplier can be used to obtain the successive terms of this sequence? Write the nth term of the sequence.

Investigation 5.10 Write the length of each line segment at stages 1, 2 and 3, if the side of the equilateral triangle at stage 0 measures 1 unit. Extend this to stages 4 and 5. Which constant multiplier can be used to obtain the successive terms of this sequence? Write the nth term of the sequence.

Investigation 5.11 Find the perimeter of the snowflake up to stage 5. Write the recursive and explicit formulae for the sequence of perimeters.

You may have observed that the number of line segments and the length of each segment at various stages of the snowflake construction leads to the geometric sequences $3, 3 \times 4, 3 \times 4^2, 3 \times 4^3, \ldots$ and $1, 1/3, 1/3^2, 1/3^3, \ldots$ respectively.

The perimeter at each stage may be obtained by multiplying the number of line segments with the length of each segment at that stage. This leads us to the following geometric sequence $3 \times 1, 3 \times 4/3, 3 \times (4/3)^2, 3 \times (4/3)^3, \ldots$.

Area enclosed by the Koch snowflake

Next we would like to find the area enclosed by the snowflake at various stages. Suppose the length of the side of the equilateral triangle at stage 0 is 1 unit. The area at stage 0 is $\frac{1}{2} \times 1 \times \frac{\sqrt{3}}{2} = \sqrt{3}/4$. At stage 1, three smaller equilateral triangles of side 1/3 units are added. The total area of these triangles is $3 \times \frac{1}{9} \times \frac{\sqrt{3}}{4} = \sqrt{3}/12$. Thus, the total area at stage 1 is $\frac{\sqrt{3}}{4} + \frac{\sqrt{3}}{12} = \sqrt{3}/3$.

Investigation 5.12 Compute the total area at stages 2 and 3. Does the sequence of total areas at the various stages form a geometric sequence? Will the total area go on increasing as n, the number of stages, approaches infinity?

You may have noted that the sequence of total areas $\frac{\sqrt{3}}{4}, \frac{\sqrt{3}}{3}, \frac{10\sqrt{3}}{27}, \ldots$ is not a geometric sequence. On the other hand, if we consider the area added at each stage, we obtain

$$\frac{\sqrt{3}}{12}, \frac{\sqrt{3}}{12} \times \frac{4}{9}, \frac{\sqrt{3}}{12} \times \frac{4^2}{9^2}, \ldots$$

which has a multiplying factor of 4/9. Observe that the total area at each stage of the Koch snowflake may be written in the form of a series as

$$\frac{\sqrt{3}}{4} + \frac{\sqrt{3}}{12}\left(1 + \frac{4}{9} + \frac{4^2}{9^2} + \cdots\right) = \frac{2\sqrt{3}}{5}$$

We can view the stages of the Koch snowflake as leading to a final shape which is simply the union of the enclosed portion at each stage. This final shape has a combination of properties which seems to defy our intuition—its area is finite but its perimeter is infinite!

Numerical and Graphical Exploration Using a Spreadsheet

To help visualise the process further, we shall explore the various attributes of the snowflake on `MS Excel`. Using our knowledge of the recursive and explicit formulae for the attributes such as length and number of line segments at each stage, perimeter and area, we shall generate values of these attributes up to stage 20 using the following steps.

Step 1: To represent the stages of the snowflake construction, generate the numbers from 0 to 20 in column A. To do this, enter 0 in cell `A2` and the formula = `A2 + 1` in cell `A3`. Drag `A3` till `A22` to obtain the list of numbers.

Step 2: In column B we shall generate the number of line segments for every stage of the snowflake (up to stage 20). For this we enter 3 in cell `B2` and the formula = `B2*4` in cell `B3`. Double click the cursor in the corner of the cell `B3` to generate the required sequence.

Step 3: To generate the sequence of the length of line segments at each stage in column C, enter 1/3 in cell `C2` and the formula = `C2*(1/3)` in cell `C3`. Double click the cursor in the corner of the cell `C3` to obtain the required sequence.

Step 4: The perimeter at every stage of the snowflake may be obtained by multiplying the number of line segments at each stage with the length of each line segment at that stage. Enter = `B2*C2` in `D2` and double click to obtain the sequence of perimeters.

Step 5: In column E, we will generate the sequence of additional areas at every stage. This will enable us to compute the total area enclosed by the snowflake at every stage. Firstly compute the value of $\sqrt{3}/4$ separately, (say in cell `H1`) by entering = `SQRT(3)/4`. Next enter 0 in cell `E2` (since there is no additional area in stage 0). In cell `E1`, we enter = `$H$1*B2*C3^2`. Double clicking this cell will lead to the sequence of additional areas at each stage of the snowflake.

Step 6: To obtain the total areas, enter = `$H$1` in cell `F2` and = `F2 + E3` in `F3`. Double clicking on `F3` will lead to the sequence of total areas.

Figure 5.2.4 shows a screenshot of the `Excel` computation. Observe that the total area of the Koch snowflake at stage 20 is equal to 0.69. This is a numerical verification of the result we discussed earlier.

A	B	C	D	E	F	G	H
stage	no. of line segments	length of line segments	perimeter	Additional area	Total area		0.433012702
0	3	1	3	0	0.433012702		
1	12	0.333333333	4	0.144337567	0.577350269		
2	48	0.111111111	5.333333333	0.06415003	0.641500299		
3	192	0.037037037	7.111111111	0.028511124	0.670011424		
4	768	0.012345679	9.481481481	0.012671611	0.682683034		
5	3072	0.004115226	12.64197531	0.005631827	0.688314861		
6	12288	0.001371742	16.85596708	0.002503034	0.690817896		
7	49152	0.000457247	22.47462277	0.00111246	0.691930355		
8	196608	0.000152416	29.96616369	0.000494427	0.692424782		
9	786432	5.08053E-05	39.95488493	0.000219745	0.692644527		
10	3145728	1.69351E-05	53.2731799	9.76645E-05	0.692742191		
11	12582912	5.64503E-06	71.03090654	4.34064E-05	0.692785598		
12	50331648	1.88168E-06	94.70787538	1.92918E-05	0.69280489		
13	201326592	6.27225E-07	126.2771672	8.57411E-06	0.692813464		
14	805306368	2.09075E-07	168.3695562	3.81072E-06	0.692817274		
15	3221225472	6.96917E-08	224.4927416	1.69365E-06	0.692818968		
16	12884901888	2.32306E-08	299.3236555	7.52734E-07	0.692819721		
17	51539607552	7.74352E-09	399.0982074	3.34549E-07	0.692820055		
18	2.06158E+11	2.58117E-09	532.1309432	1.48688E-07	0.692820204		
19	8.24634E+11	8.60392E-10	709.5079242	6.60837E-08	0.69282027		
20	3.29853E+12	2.86797E-10	946.0105656	2.93705E-08	0.6928203		

Fig. 5.2.4 Computing the values of the various attributes in stages 0 to 20 of the Koch snowflake

5.3 SPACE FILLING TREES

In this section, we shall explore some special kinds of fractals known as space filling trees. These trees are self-similar in nature and are generated using recursive construction processes just like the other fractals you have explored in this chapter.

To understand the process, we shall construct a space filling tree inside an equilateral triangle. We shall use a skeletal triangular grid (Fig. 5.3.1), as a reference. In each successive stage we join the midpoints of the sides of the smaller triangles to obtain new subtriangles. Thus, at stage 0 we start with one triangle, at stage one there are 4 subtriangles, at stage two there are 16 and so on. Thus, at the nth stage there are 4^n subtriangles.

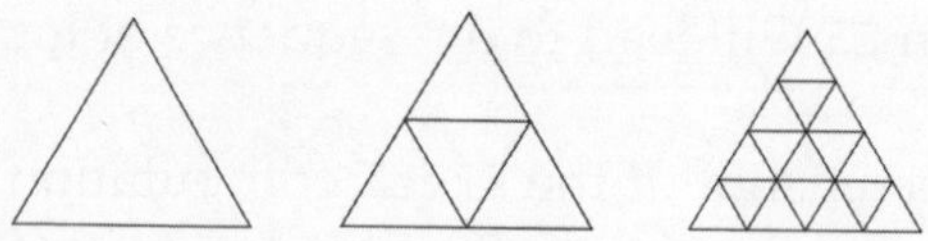

Fig. 5.3.1 The skeletal triangular grid—stages 0, 1 and 2

We shall first evolve a method of labelling the subtriangles at each stage as follows. Note that stage 1 comprises four smaller subtriangles. Let us label the centre triangle with C, the top triangle with T, the left triangle with L and the right triangle with R. Figure 5.3.2 shows the labelling.

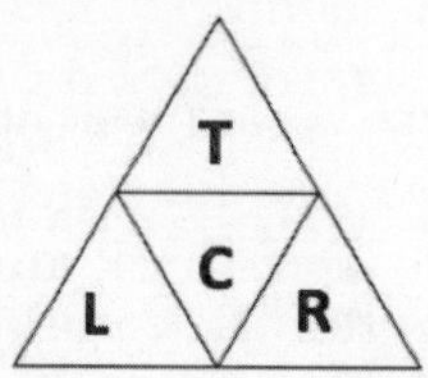

Fig. 5.3.2 Labelling the subtriangles in stage 1

Now let us try to label the sixteen triangles of stage 2. Here we have to imagine that all the four subtriangles C, T, L and R are subdivided into four smaller equilateral triangles. First, consider the top triangle T of stage 1. This comprises four smaller subtriangles. We label these as TC, TT, TL and TR. Similarly we label the subtriangles of L and R as LC, LT, LL, LR, RC, RT, RL and RR respectively. The triangle C of stage 1 however has a

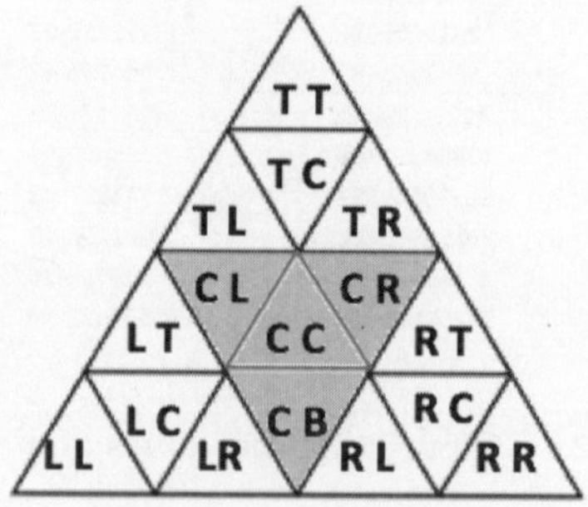

Fig. 5.3.3 Labelling the subtriangles in stage 2

different orientation. It is 'upside down' as compared to triangles T, L and R. Thus we will label its subtriangles as CC, CL, CR and CB (Fig. 5.3.3).

Now that we have decided on the labelling process, we are ready to start the construction of our space filling tree. Consider stage 1 of the triangular grid (Fig. 5.3.2) which comprises 4 smaller subtriangles T, L, R and C. Starting with the centre of C (let us label it as X_C) we draw three line segments, each ending at one of the centres of the three outer subtriangles T, L and R. Let us label these centres as X_T, X_L and X_R respectively. The first iteration of the required tree thus comprises the three line segments $X_C X_T$, $X_C X_L$ and $X_C X_R$ (see Fig. 5.3.4). We shall refer to this stage as T_1 (see Fig. 5.3.5).

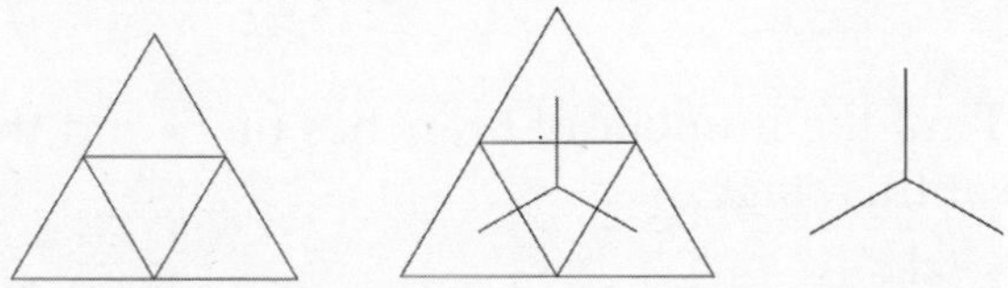

Fig. 5.3.4 The first iteration of the space filling tree

Once the first iteration is obtained, we remove the triangular grid from the background. If we look at the vertices or endpoints of the tree we observe that vertex X_C has three branches coming out of it. In the language of graph theory, we say that X_C has degree 3. In a similar manner we observe that vertices X_T, X_L and X_R have degree 1 each.

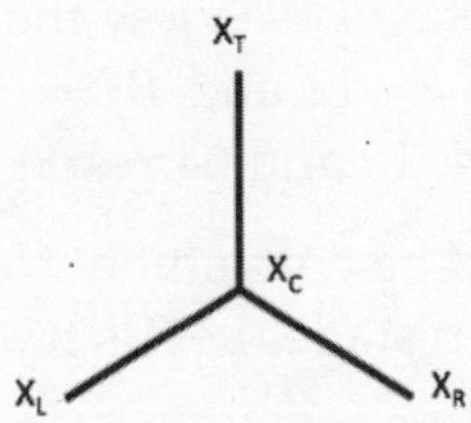

Fig. 5.3.5 T_1 or stage 1 of the space filling tree

For generating iteration 2 of the tree, namely T_2, we shall refer to the sixteen subtriangles of stage 2 of the triangular grid. Starting with the centres of the four subtriangles T, L R and C of stage 1 (presently labelled as X_T, X_L, X_R and X_C) we draw line segments to each of the centres of the subtriangles TT, TL, TR, LT, LL, LR, RT, RL, RR, CL, CR and CB. We now obtain 12 smaller branches

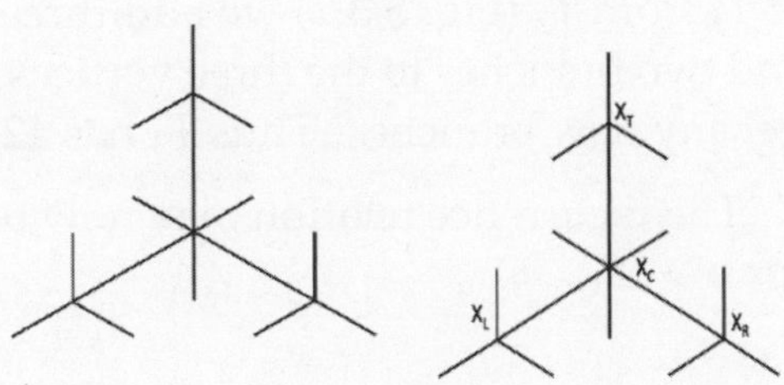

Fig. 5.3.6 Stage 2, or T_2, of the space filling tree

added to the initial three line segments. Once this is done, we remove the triangular grid from the background. This iteration is shown in Fig. 5.3.6.

Thus T_2 has 12 branches. Also there are 12 vertices of degree 1 (these are the terminal vertices), 3 vertices of degree 4 (X_T, X_L, X_R) and 1 vertex of degree 6 (X_C).

Let us now try to compute the total length of the tree in stage 2. Let us suppose that the original equilateral triangle has sides of unit length. Then each branch of the first iteration is of length 1/3. Thus, the sum of the lengths of the branches in the first iteration is $X_CX_T + X_CX_L + X_CX_R = 1/3 + 1/3 + 1/3 = 1$.

In T_2, 12 new branches get added where each branch is half the length of the branches in T_1, that is, 1/6. Thus the total length of the branches in T_2 is $3 \times \frac{1}{3} + 12 \times \frac{1}{6} = 3$.

Investigation 5.13 Find the number of branches of T_3 and the total length of the branches of the tree at this stage.

Counting Branches

We may now describe our space filling tree in terms of the number of branches B_n at the nth stage. We view the branching as starting from the T_1 stage. Thus $B_0 = 0$, $B_1 = 3$, $B_2 = 12$ and $B_3 = 42$. Also the number of vertices at the T_1, T_2 and T_3 stages are 4, 16 and 64 respectively. Thus the number of vertices in T_n is 4^n.

It would be convenient if we had a formula for finding the number of branches at any given stage. Let us try to find a recurrence relation expressing B_{n+1} in terms of B_n.

First we observe that every stage of the tree (T_2 onwards) has vertices of degree 1, 4 and 6 only. We shall refer to the vertices of degree 1 as terminal vertices. At any stage, three branches get added to each terminal vertex of the previous stage. Also two branches get added to all the vertices of degree 4 of the previous stage while the vertices of degree 6 do not get any additional branches. Thus, to arrive at T_3 from T_2 (Fig. 5.3.7) we add three branches to each of the 12 terminal vertices and two branches to the three vertices of degree 4 (namely X_T, X_L, X_R). X_C doesn't get any new branches. Thus T_3 has 42 branches in all.

The recurrence relation may now be written as $B_{n+1} = 3B_n + 2(4^n - B_n - 4^{n-2})$ for $n > 2$.

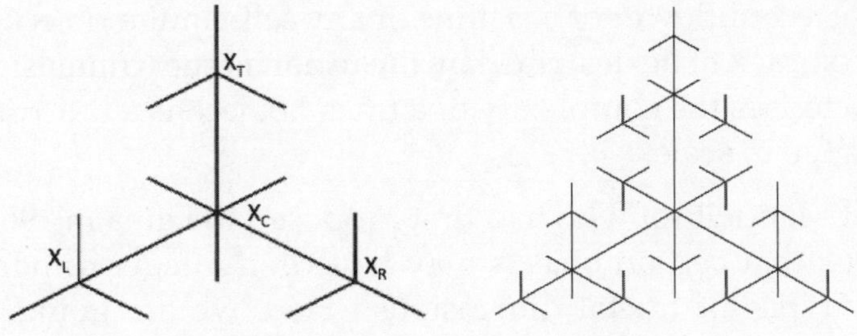

Fig. 5.3.7 Generating T_3 from T_2

At stage $n+1$, three branches get added to the terminal vertices of T_n, and 2 branches get added to the vertices of degree 4 which are $(4^n - B_n - 4^{n-2})$ in number.

Investigation 5.14 Solve the recurrence relation $B_{n+1} = 3B_n + 2(4^n - B_n - 4^{n-2})$ for $n > 2$ to find a formula for the total number of branches at stage n.

The reader may be wondering why the tree is referred to as a space filling tree. Well, it actually fills the equilateral triangle inside which we have constructed it, in the sense that it enters any disc placed over the triangle, regardless of how small the disc is. This may not be evident in the initial stages but as the stages progress, we see how the filling occurs. Figure 5.3.8 shows the higher stages of our space filling tree.

Fig. 5.3.8 Stages T_5 and T_6 of the space filling tree. Image credit [C9].

5.4 Fractal Dimension

In this section, we shall explore an important aspect of fractal objects—fractal dimension. In Euclidean geometry, we have learnt about the concept of dimension and have been accustomed to the fact that a point is dimensionless, a line has dimension equal to one, polygonal shapes drawn on a planar surface are two dimensional and objects in space are three dimensional. But how do we describe the dimension of a cauliflower or coastline or any self-similar object. In fact, fractal objects are too complex to be described by the usual integer dimensions. The *fractal dimension* characterises the complexity of a fractal object as a ratio of the change in detail to the change in scale.

It was Benoit Mandelbrot who had first proposed the idea in 1967, in his paper on self-similarity, that certain objects may have non-integer dimensions. In fact there are many types of fractal dimensions. Here we are going to discuss the notion of self-similarity dimension specifically with reference to the Sierpinski triangle, Sierpinski carpet and Koch snowflake which we have already explored earlier in this chapter.

The rule we will use to measure self-similarity dimension is $a = 1/s^D$ where a denotes the number of parts, s the scaling factor and D the dimension.

Consider a straight-line segment of unit length. If we divide it into two equal parts, each part will be of length $1/2$ units. Thus, in this case $a = 2$ (as the line segment is divided into two equal parts) and $s = 1/2$ (each part has been scaled down to half the original length). Substituting in $a = 1/s^D$ we get $2 = 1/(1/2)^D$ which simplifies to $2 = 2^D$. Hence $D = 1$ which indicates that a line segment is one-dimensional.

What happens if we divide the line segment into three equal parts? There will be three equal parts and each part will be of length $1/3$ unit. Substituting in our formula we get $3 = 1/(1/3)^D$ which simplifies to $3 = 3^D$. Once again we get $D = 1$ and this will be the case no matter how many parts we divide the original line segment into. Let us now consider a unit square and mark the midpoints of the four sides. When we join the midpoints of the opposite sides we get four smaller squares, each of side $1/2$ unit. Using our formula and substituting $a = 4$, $s = 1/2$ we get $4 = 1/(1/2)^D$ which simplifies to $D = 2$ which indicates that a square is two-dimensional. Similarly if we had trisected the sides of the unit square and joined the points of trisection of the opposite sides, we would obtain 9 smaller squares of side $1/3$ unit each. The formula would now yield $9 = 1/(1/3)^D$ which again gives $D = 2$ and this will be preserved no matter how many parts we divide the square into.

Investigation 5.15 Use the formula $a = 1/s^D$ to justify that a cube is three dimensional.

Let us now use the formula to compute the dimension of the Sierpinski triangle. At stage 1, we have three copies of stage 0 and the side length of each smaller equilateral triangle is $1/2$ unit (assuming that the side of the original triangle at stage 0 is 1 unit). Thus $a = 3$ and $s = 1/2$. Substituting in $a = 1/s^D$ we get

$3 = 1/(1/2)^D$ or $3 = 2^D$. In order to compute D we need to apply log on both sides of the equation: $\log 3 = D \log 2$ implies $D = \log 3/\log 2 \approx 1.585$. Let us now consider stage 2 of the fractal. Here we have 9 smaller equilateral triangles each of side length 1/4 unit. Substituting $a = 9$ and $s = 1/4$ we get $9 = 1/(1/4)^D$, which leads to $\log 9 = D \log 4$ or $D = \log 9/\log 4 = 2\log 3/2\log 2 = \log 3/\log 2$. Thus the self-similarity dimension of the Sierpinski triangle is approximately 1.585.

Investigation 5.16 Using the reasoning given above, compute the self-similarity dimension of the Sierpinski carpet.

In the case of the Sierpinski triangle and Sierpinski carpet, it is easy to see the number of copies of the previous stages in a given stage. This enables us to find the self-similarity dimension. Can the same approach be used to find the self-similarity dimension of the Koch snowflake? Notice that to obtain stage 1 of the snowflake, each side of the equilateral triangle (at stage 0) is replaced by four identical line segments of length 1/3. Thus substituting $a = 4$ and $s = 1/3$ in $a = 1/s^D$ we get $4 = 1/(1/3)^D$ or $4 = 3^D$. Applying log on both sides leads to $D \log 3 = \log 4 = 2 \log 2$, which implies $D = 2\log 2/\log 3 \approx 1.26$.

It is possible to find the self-similarity dimension of most fractals using this approach.

FURTHER EXPLORATIONS

In this chapter an attempt has been made to highlight the basic properties of some geometrical fractals such as the Sierpinski triangle, Sierpinski carpet, Koch snowflake and a space-filling tree. There are many fractals which can lead to interesting investigations. Two such fractals that are worth mentioning are the Pythagorean tree [86] and the Koch anti-snowflake [82]. The emphasis in this chapter has been on exploring various attributes of fractals at different stages, such as number of line segments, perimeter, area and so on using `Excel`. `Excel` helped us to observe the patterns numerically and gave us an insight into the iterative processes which generate these fractals. However these fractals are hard to generate pictorially. The article [17] explains how a dynamic geometry software such as `GeoGebra` may be used to generate various stages of the Sierpinski triangle and the Pythagorean tree. In the last section of the chapter, we have described the notion of self-similarity dimension of fractals objects and have used it to compute the fractal dimension of the Sierpinski triangle and the Sierpinski carpet. There are other methods of exploring fractal dimension. Two such methods are the Sandbox method and the Box Counting method which may be found in [5].

6

The Newton–Raphson Method and Fractals

In the previous chapter, we encountered and studied shapes that had self-similarity. These 'fractals' were created by repeated geometric moves. In this chapter we shall create further fractal shapes, except these will emerge from repeated *algebraic* moves. The mathematics of this chapter is a little more advanced and involves some basic calculus. Specifically, we need the notion of the derivative of a function. We shall also use complex numbers. On the programming side, we'll work with `Mathematica`.

We begin with the practical problem of solving a given equation. You have learned how to solve linear and quadratic equations. These have the forms

Linear: $ax + b = 0$

Quadratic: $ax^2 + bx + c = 0$

with $a \neq 0$. The coefficients a, b, c are real numbers. A linear equation always has the real solution $x = -b/a$. However, the quadratic equation may fail to have real solutions, although it will at least have complex solutions. The method for solving a quadratic equation when its solutions are real has been known for some 4000 years, dating back to ancient Egypt and Mesopotamia!

Next, we may wonder about equations that involve polynomials of higher degrees. A 'cubic equation' involves a third-degree polynomial while a 'quartic polynomial' involves a fourth-degree polynomial. These are intrinsically harder, in the sense that the use of complex numbers is unavoidable, even when the solutions are real numbers. Their general solutions were found only in the 16th century by a group of Italian mathematicians. The formulae are quite complicated. For example, we can ask `Mathematica` to solve the innocent-looking equation $x^3 + 3x + 1 = 0$ by the following command:

```
Solve[x^3-3*x+1==0, x, Cubics->True]
```

The resulting solutions are quite startling, and demonstrate why you were not asked to solve general cubic equations in school.

$$\left\{\left\{x \to \frac{1}{(\frac{1}{2}(-1+i\sqrt{3}))^{1/3}} + (\frac{1}{2}(-1+i\sqrt{3}))^{1/3}\right\}\right.$$
$$\left\{x \to -\frac{1-i\sqrt{3}}{2^{2/3}(-1+i\sqrt{3})^{1/3}} - \frac{1}{2}\left(\frac{1}{2}(-1+i\sqrt{3})\right)^{1/3}(1+i\sqrt{3})\right\}$$
$$\left.\left\{x \to -\frac{1}{2}(1-i\sqrt{3})\left(\frac{1}{2}(-1+i\sqrt{3})\right)^{1/3} - \frac{1+i\sqrt{3}}{2^{2/3}(-1+i\sqrt{3})^{1/3}}\right\}\right\}$$

A remarkable fact is that there are no general formulae for equations involving polynomials of degree five or higher. It is not that we have failed to find these solutions due to insufficient cleverness on our part. It was proved in the 19th century that *there is no general method* of solving these equations using only the arithmetic operations and roots! So what do we do? The next best thing is to come up with ways of finding approximate solutions. In this chapter, we shall study one such method and the further issues that it generates.

The first steps in this area were taken by Arthur Cayley (1821–95), when he studied a process of successive approximations to a root of a quadratic equation over complex numbers and was able to describe how the process depended on the choice of the first approximation. He wished to do the same for cubic equations but was not able to. As you read this chapter, you will understand why. The next steps were taken by Pierre Fatou (1878–1929) and Gaston Julia (1893–1978). Fatou and Julia applied new developments in complex analysis to this problem and identified the basic features. Their achievements seem even more astonishing with the passing of time, as computers reveal how intricate the situation is. As noted in the previous chapter, Benoit Mandelbrot (1924–2010) popularised this area of mathematics by demonstrating its applicability to various natural and economic processes. In 1998, Curt McMullen received the Fields Medal, the highest prize in Mathematics, for his success in extending the theory and identifying new properties of the sets being studied. For more information on the life and work of these mathematicians, you can visit the MacTutor History of Mathematics Archive —Cayley [42], Fatou [47], Julia [50], Mandelbrot [54] and McMullen [55].

6.1 The Newton–Raphson Method for Real Functions

Let us investigate $p(x) = x^3 - 3x + 1$ as a function of real numbers, by plotting its graph. Since we'll be using it repeatedly, we first set up its definition.

```
p[x_] := x^3-3*x+1;
Plot[p[x], {x,-3,2}, PlotRange->{-2, 4}]
```

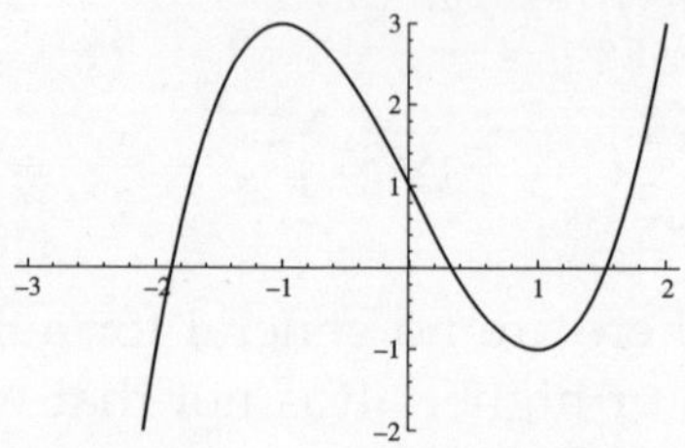

Fig. 6.1.1 A Mathematica drawn plot of $x^3 - 3x + 1$

Figure 6.1.1 shows the locations of the three roots of $p(x)$. They are all real, in spite of the complex expressions provided by Mathematica. This is a quirk of cubic equations and not the fault of Mathematica. Now, drawing this picture involves a good amount of computation in the background. Can we use it to motivate an approach which has few calculations? A simple idea is to choose a point on the graph and move tangentially from it towards the x-axis. If the tangent does not deviate too much from the graph then they will cross the axis at nearby points. Recall that the equation of the tangent line at $(a, p(a))$ is $y = p(a) + p'(a)(x - a)$. For example, we take $a = -1.5$ as a starting point. The following command plots the function and the tangent line together.

```
Plot[{p[x], p[-1.5]+p'[-1.5]*(x+1.5)}, {x,-3,2}, PlotRange->{-2, 4}]
```

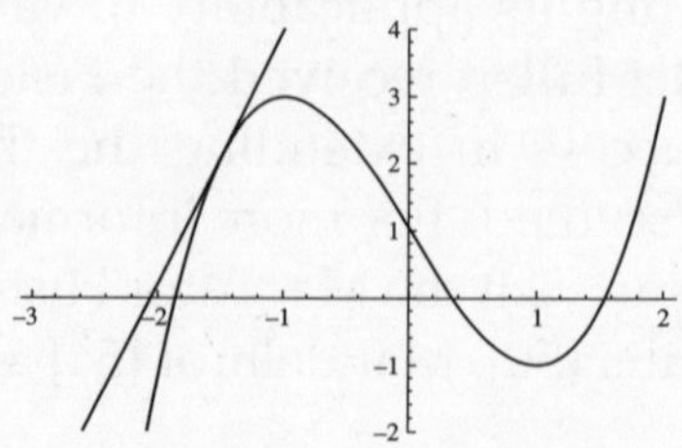

Fig. 6.1.2 Graph of $x^3 - 3x + 1$ and a tangent line

The tangent line does cross the x-axis not too far from the sought solution. We can solve for the crossing point with the code

```
NSolve[p[-1.5]+p'[-1.5]*(x+1.5)==0, x]
```

to get $x = -2.067$. We repeat the process to improve it further, and

```
NSolve[p[-2.067]+p'[-2.067]*(x+2.067)==0, x]
```

gives $x = -1.901$. We can continue.

Investigation 6.1 Verify that iterating the above procedure gives the following sequence of numbers: -1.87972, -1.87939, -1.87939.

The process appears to stabilise near $x = -1.87939$. Indeed, we find that $p(-1.87939) = -0.000036$ is quite close to zero.

The method we have described is known as the Newton–Raphson method. We can summarise it as follows. Start with any value x_0 of x. Solve the equation $p(x_0) + p'(x_0)(x - x_0) = 0$ to get a new value $x_1 = x_0 - p(x_0)/p'(x_0)$. Repeat the process with x_1 to get $x_2 = x_1 - p(x_1)/p'(x_1)$. Continuing in this way, obtain points x_3, x_4, and so on. The sequence x_0, x_1, x_2, ... generated in this manner is called the 'orbit' of x_0.

The Newton–Raphson method comes with a guarantee that if you start with x_0 which is close to a solution $x = c$ of $f(x) = 0$, where f is a polynomial, then $x_1 = x_0 - f(x_0)/f'(x_0)$ is closer still. To justify this, we begin with the observation that $f(x) = (x - c)p(x)$ for some polynomial $p(x)$. Then, if $p(c) \neq 0$,

$$\begin{aligned} x_1 - c = x_0 - c - \frac{f(x_0)}{f'(x_0)} &= x_0 - c - \frac{(x_0 - c)p(x_0)}{p(x_0) + (x_0 - c)p'(x_1)} \\ = (x_0 - c)^2 \frac{p'(x_0)}{p(x_0) + (x_0 - c)p'(x_0)} &\approx (x_0 - c)^2 \frac{p'(c)}{p(c)} \end{aligned}$$

If x_0 is close enough to c then $|x_0 - c||p'(c)/p(c)| < 1$ and $|x_1 - c| < |x_0 - c|$.

The fun begins when we ask about what happens if we start well away from a solution, and we shall now explore this numerically. Clearly, it would be helpful to define the function that is being repeatedly applied:

```
nr[x_] := x-p[x]/p'[x]
```

We can repeatedly apply the Newton–Raphson step to generate points of an orbit as follows:

```
x[1]= -1.5;
iterations= 7;
For[k=1, k<iterations, k++, x[k+1]= N[nr[x[k]]]];
Table[x[k], {k, 1, iterations}]
```

Investigation 6.2 Explore what happens if the Newton–Raphson process is initiated from other starting points such as 0, ± 1, ± 2, ± 1000.

In this example, we see that different starting points take us to different solutions. This raises the following questions:

(i) Will every starting point always lead to some solution?

(ii) Can we identify which starting points will lead to a given solution?

Investigation 6.3 Apply the Newton–Raphson process to $x^3 - 2x + 2 = 0$ with the starting points $x_0 = 0, 1, 5$ and 10.

Investigation 6.4 Apply the Newton–Raphson process to $\arctan(x) = 0$ with starting points near $x_0 = 1.39175$.

These investigations provide examples of starting points that fail to lead to a solution. Sometimes the points grow in magnitude and move towards the ends of the number line instead of settling down to one value. At other times, they settle into a repetitive pattern and again fail to settle down to a solution. We also see that a starting point need not lead to the closest solution. For example, if we start the Newton–Raphson process for $x^3 - 3x + 1 = 0$ at $x = -0.8$, we are led to the farther off solution $x = 1.532$ rather than the nearest solutions at $x = 0.347$ and $x = -1.879$. Let us explore these possibilities with a polynomial which has easy calculations: $q(x) = x^3 - x$. Its roots are 0 and ± 1, while the critical points are at $\pm 1/\sqrt{3}$. The following code generates the graph and the critical points:

```
q[x_]:= x^3-x;
qplot= Plot[q[x], {x, -1.5, 1.5}];
critical= x/.NSolve[q'[x]==0, x];
qpoints= ListPlot[Table[{critical[[k]], q[critical[[k]]]}, {k, 1,
2}]];
Show[qplot, qpoints]
```

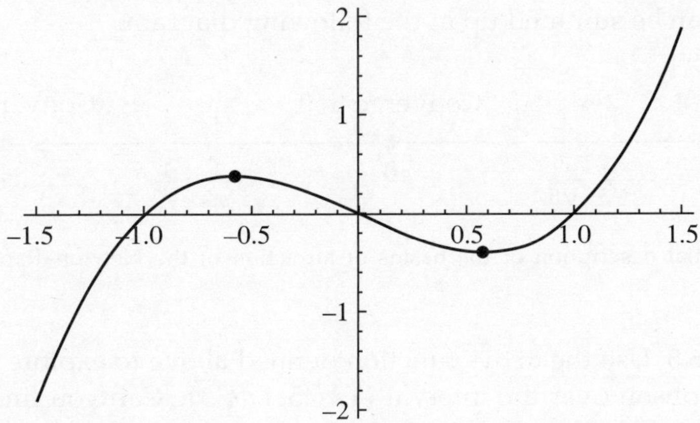

Fig. 6.1.3 Graph of $q(x) = x^3 - x$ showing the critical points at $\pm 1/\sqrt{3}$

For each root, the corresponding 'basin of attraction' is the collection of starting points which converge to it under the Newton–Raphson process. Figure 6.1.3 clearly indicates that the interval $[1/\sqrt{3}, \infty)$ will be in the basin of attraction of $x = 1$ while $(-\infty, -1/\sqrt{3}]$ will be in the basis of attraction of $x = -1$. The nature of points between $\pm 1/\sqrt{3} = \pm 0.57735$ is less clear, although the ones near 0 should be in its basis of attraction. But how near is enough?

The following code defines a function which identifies the likely limit of the Newton–Raphson method applied to a point x_0. It stops when the magnitude of the function value drops below 0.001 or else after 100 iterations.

```
iterations=100;
orbit[x0_]:= Module[{x=x0,k=1},
While[Abs[q[x]]>0.001 || k<iterations,{x,k}={nrq[x],k+1}];
x]
```

Investigation 6.5 Collect data on the basins of attraction for $x^3 - x$ by applying the Newton–Raphson method to the starting points $0, 0.2, 0.4, \ldots, 1.4$. Do we need separate calculations for negative starting points?

Investigation 6.6 Show that the starting points $\pm 1/\sqrt{5} = \pm 0.4472$ form a '2-cycle', that is, Newton–Raphson moves from one of these to the other and back. Show this is the only 2-cycle for $x^3 - x$. (Use `Mathematica` for the last part.)

Investigation 6.7 Let $n(x) = x - q(x)/q'(x)$. Show that $|x| < 1/\sqrt{5}$ implies $|n(x)| < |x|$. Hence the points in $(-1/\sqrt{5}, 1/\sqrt{5})$ are in the basin of attraction of 0.

Our work can be summed up in the following diagram.

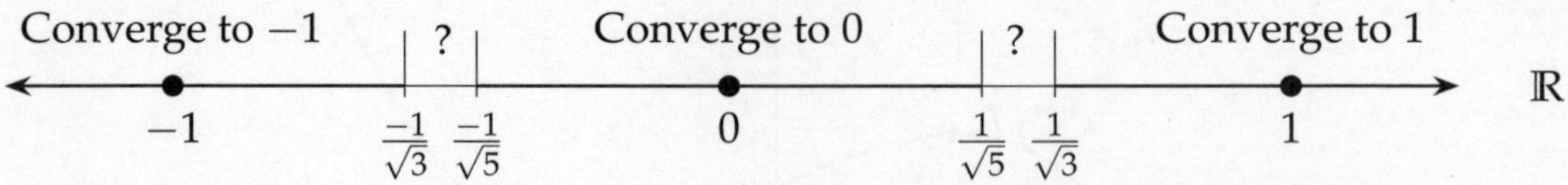

Fig. 6.1.4 A partial description of the basins of attraction of the Newton–Raphson process for $q(x) = x^3 - x$

Investigation 6.8 Use the `orbit` function defined above to explore the behaviour of Newton–Raphson over the interval $(1/\sqrt{5}, 1/\sqrt{3})$. Can you find subintervals which are in certain basins of attraction?

One of the truths revealed by these calculations is that tiny deviations in the starting point of an iterative process can have a drastic impact on the final result. The details can be explored numerically but are hard to visualise when we only have a line to work with. More interesting phenomena are revealed when we work with complex numbers and have a whole plane in which to wander.

6.2 The Newton–Raphson Method for Complex Functions

Consider a function $f(z)$ whose input and output are complex numbers. Its derivative can be defined just as for a function of real numbers:

$$f'(z) = \lim_{h \to 0} \frac{f(z+h) - f(z)}{h}$$

where h takes complex values and $h \to 0$ means that the magnitude of h goes to zero. This complex derivative has the same basic rules as the real derivative. In particular $1' = 0$ and $(z^n)' = nz^{n-1}$ for $n = 1, 2, 3, \ldots$. Solutions of $f(z) = 0$ can again be sought using the Newton–Raphson method. Start with some complex number z_0 and generate others using the rule

$$z_{n+1} = z_n - \frac{f(z_n)}{f'(z_n)}$$

Let us again begin with a simple example, the equation $z^2 - 1 = 0$. This has two solutions: $z = \pm 1$. The Newton–Raphson iteration is

$$z_{n+1} = z_n - \frac{z_n^2 - 1}{2z_n} = \frac{1}{2}\left(z_n + \frac{1}{z_n}\right) \tag{6.7}$$

The following diagram shows five orbits created by applying this iteration to the starting points $2 \pm i, 1-i, 0.2-i$ and $0.5+i$. All converge to $z = 1$.

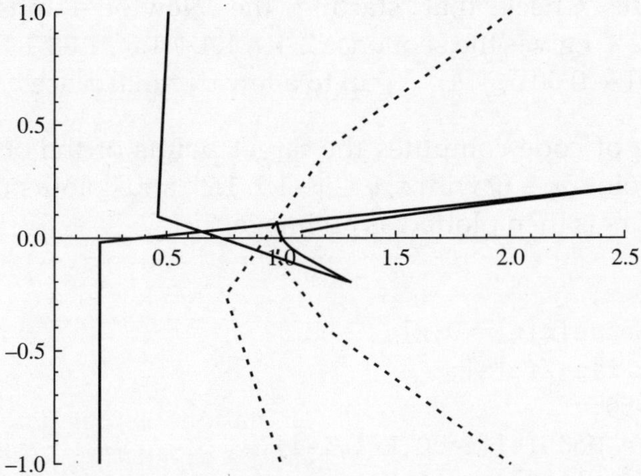

Fig. 6.2.1 Convergence of orbits of the Newton–Raphson iteration for $z^2 - 1 = 0$

Figure 6.2.1 was produced by the following code.

```
initial={2.+I,2.-I,1.-I,.2-I,.5+I};
For[j=1,j<6,j++,
 Module[{},
  x[1,j]=initial[[j]];
  For[k=1,k<5,k++,x[k+1,j]=(x[k,j]+1/x[k,j])/2];
  steps[j]=Table[N[x[k,j]],{k,1,5}]
]];
ComplexListPlot[Table[steps[j],{j,1,5}],Joined->True,
 PlotStyle->{{Dotted,Black},{Dotted,Black},{Dashed,GrayLevel[.3]},
  Black,GrayLevel[.4]},
 PlotRange->{{0,2.5},{-1,1}}]
```

Investigation 6.9 Show that points on the imaginary axis stay on that axis under the iteration (6.7). Use `Mathematica` to explore the basins of attraction of ± 1.

We see that points with positive real part are attracted to 1, while those with negative real part are attracted to -1. These two basins of attraction are separated by the imaginary axis.

Next, consider $f(z) = z^3 - z$. We already know its three solutions $z = 0, \pm 1$. We have some idea of what happens if we start the Newton–Raphson method from a real number. The `Mathematica` code developed earlier can be used for our explorations over complex numbers as well.

Investigation 6.10 Check that starting the Newton–Raphson method for $z^3 - z = 0$ at $z = 2 + i$ gives the sequence $2 + i$, $1.4 + 1.6i$, $1.08 + 0.33i$, $0.94 + 0.33i$, $0.98 - 0.02i$, $1.0001 + 0.001i$, $1, 1, \ldots$, up to a few decimal places.

The next piece of code computes the target points of the orbits arising from various starting points $x + iy$ with $x, y \in [-1.2, 1.2]$, and colours them accordingly. The array of colours is then plotted as an image.

```
f[z_]:= z^3-z;
roots=z/.NSolve[f[z]==0,z];
nrf[z_]:= z-f[z]/f'[z];
iterations=50;
orbit[z0_]:= Module[{z=z0,k=1,i=1},
 While[Abs[f[z]]>0.001 && k<iterations,{z,k}={nrf[z],k+1}];
 Scan[If[Abs[z-#]<0.001,Return[i],i++]&, roots];
 If[i<=3,{i},None]];
basins=Table[orbit[x+y*I],{y,-1.2,1.2,.004},{x,-1.2,1.2,.004}];
colors= {GrayLevel[.5],GrayLevel[.9], GrayLevel[0.2]};
fractal=basins/.{{k_Integer}:>(colors[[k]]),None->GrayLevel[0]};
Graphics[RasterArray[fractal]]
```

The `GrayLevel` settings were used because this book is not in colour. You should replace them by `RGBColor[r,g,b]` with r, g, b taking values in the interval $[0, 1]$.

The above code produces the image displayed on the left below. The darker regions are the basins of attraction for ± 1 while the light region is the basin for 0. The image in the centre was obtained by zooming into one of the 'bulbs' (with $x \in [0.42, 0.62]$ and $y \in [-0.1, 0.1]$.

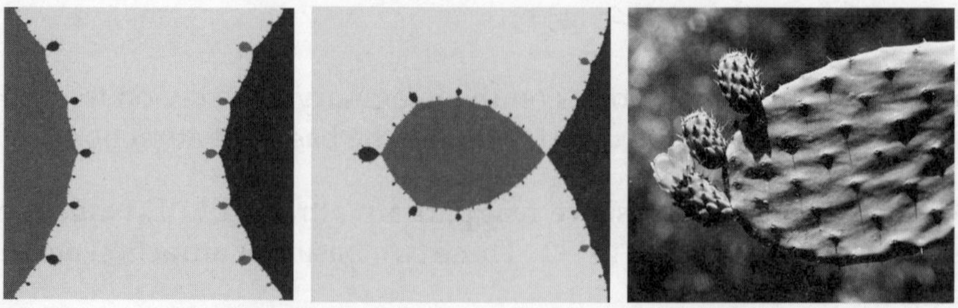

Fig. 6.2.2 Two views of the basins of attraction for $z^3 - z = 0$. On the extreme right is a photograph of a prickly pear cactus! Image credit [C10].

More interesting patterns can be obtained by varying the configuration of the zeroes of the polynomial. For example, if we take $g(z) = z^3 - 1$, the roots are placed equidistantly along the unit circle. The symmetry in their relative location leads to symmetry between the basins of attraction.

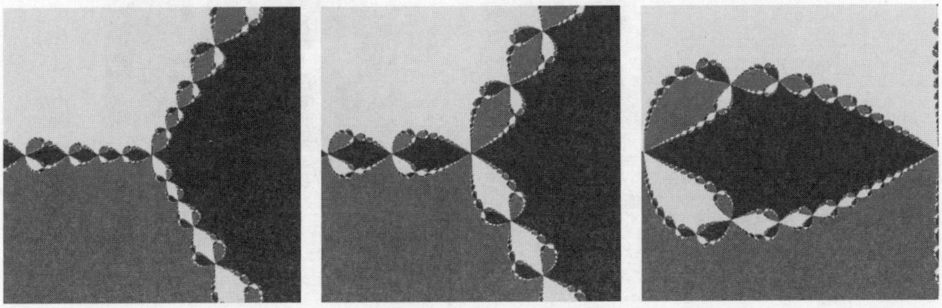

Fig. 6.2.3 The basins of attraction for $z^3 - 1 = 0$

Investigation 6.11 Produce the images in Fig. 6.2.3 using `Mathematica`.

Investigation 6.12 Use the polynomials $z^4 - 1$, $z^5 - z$ and $(8z^3 - 1)(z^2 - 1)$ to explore the fractals in Fig. 6.2.4.

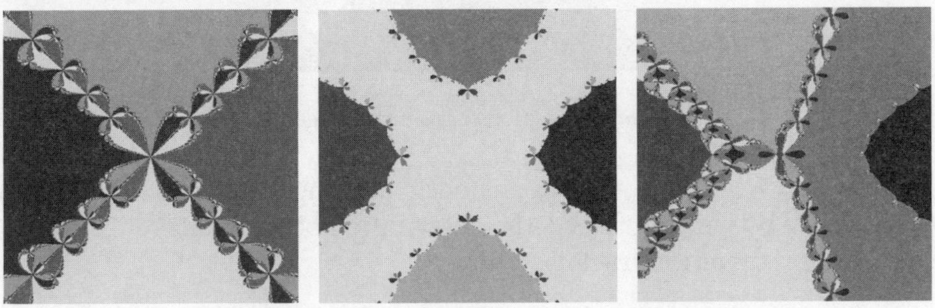

Fig. 6.2.4 A variety of fractals produced by Newton–Raphson iterations

In the examples we have considered so far, the basins of attraction essentially fill the plane. The starting points which do not lead to a solution seem to be sparse. For example, in the $z^2 = 1$ example, they are confined to the imaginary axis. In the other examples too, they form at best a sort of elaborate cobweb separating the basins of attraction. However, this is not always the case. The solid black regions in the diagrams below consist of exactly such points. These diagrams correspond to the equation $z^3 - 2.55z^2 - 1 = 0$. (See the website [84] for more details of this particular example, as well as many other intriguing examples and visualisations.)

Note how the close-up reveals a portion that looks like a Koch snowflake drawn with semicircles!

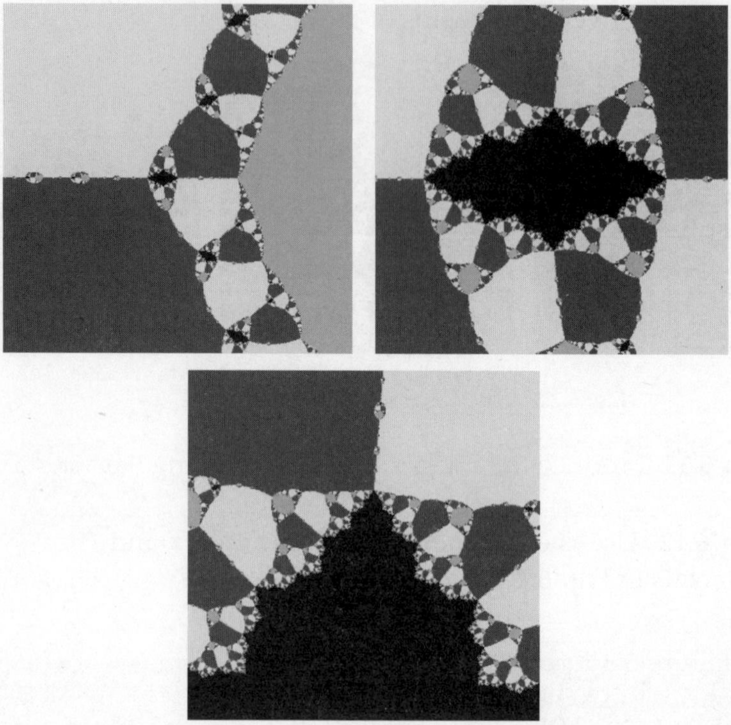

Fig. 6.2.5 Regions of non-convergence

Investigation 6.13 Explore the persistence and behaviour of the regions of non-convergence by varying the coefficients of the polynomial $z^3 - 2.55z^2 - 1$ and drawing the corresponding fractals.

In the next section, we shall take a closer look at the phenomenon of orbits that do not converge to a solution.

6.3 Cyclic and Chaotic Orbits

We had found the 2-cycle $\{\pm 1/\sqrt{5}\}$ for the Newton–Raphson method applied to $x^3 - x = 0$. Let us look for more 2-cycles. The requirement for a double application of Newton–Raphson to return to the original point creates a polynomial equation of degree 9. So it should have 9 solutions over the complex numbers. Three of these are the roots $0, \pm 1$, and we already have the 2-cycle $\{\pm 1/\sqrt{5}\}$. Therefore, there are potentially two more 2-cycles to be found. Due to the symmetry with respect to the x-axis, we hope that these may occur as conjugate pairs.

Investigation 6.14 Solve $z - \dfrac{z^3 - z}{3z^2 - 1} = \bar{z}$ to obtain $z = \dfrac{\sqrt{7}}{4} \pm \dfrac{i}{4}$ and $z = -\dfrac{\sqrt{7}}{4} \pm \dfrac{i}{4}$.

The following diagram shows the result of asking `Mathematica` to apply the Newton–Raphson step repeatedly to $1/\sqrt{5}$.

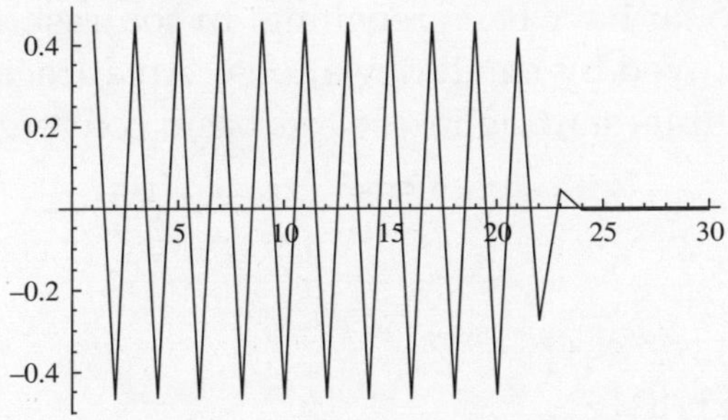

Fig. 6.3.1 A 2-cycle that degenerates and converges to a point

The cycle in Figure 6.3.1 persisted for a while but then suddenly degenerated and collapsed to one of the roots. Evidently this is because small rounding errors in the calculations moved the points away from the cycle and into a basin of attraction. The complex cycles $\{(\sqrt{7} \pm i)/4\}$ and $\{(-\sqrt{7} \pm i)/4\}$ have the same behaviour. The calculated orbits in the complex plane are shown below, with the eventual sudden movement to a root.

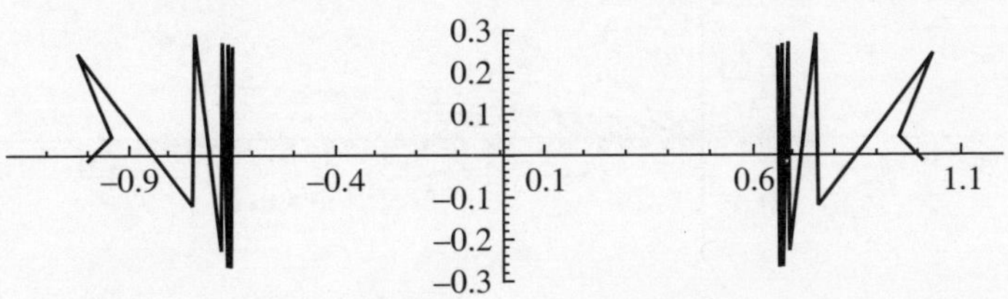

Fig. 6.3.2 2-cycles in the complex plane that degenerate and converge to a point

We can have longer cycles as well. For example, a 3-cycle has the form $z_1, z_2, z_3, z_1, \ldots$. The 3-cycles arising out of a cubic equation are solutions of a polynomial of degree 27. After discounting the three roots, we can typically expect eight 3-cycles. In this way, we have increasing numbers of cycles of higher lengths, and all the points of these cycles are examples of initial points for which Newton–Raphson does not converge to a point. Except that in numerical work, the rounding errors may destroy the cycle as we already saw.

Investigation 6.15 Use `Mathematica` to find the 3-cycles for $z^3 - z = 0$. Iterate these orbits many times to see if they are stable under rounding errors.

If small deviations destroy a cycle, we call it a 'repelling cycle'. If we are interested in using Newton–Raphson only to solve equations, then repelling cycles are good news, as they push us away into a basin of attraction of a root. All the cycles we have seen so far have been repelling. In contrast, an 'attracting cycle' is one which is not destroyed by small deviations. An attracting cycle has its own 'basin of attraction': Orbits starting inside this basin converge to the cycle.

Consider the cubic equation $z^3 - 2.55z^2 - 1 = 0$. We can find the solutions and 2-cycles as follows.

```
h[z_]:= z^3-2.55*z^2-1;
nrh[z_]:= z-h[z]/h'[z];
roots= z/.NSolve[h[z]==0, z];
cycles= z/.NSolve[nrh[nrh[z]]==z, z]
```

We find the following 2-cycles:

$$\{1.97188 \pm 0.567162i\}, \{0.889861, -0.180352\}, \{0.668217, -0.221492\}.$$

Figure 6.3.3 illustrates that if we start at a point near enough to 0.889861, we are pulled into the $\{0.889861, -0.180352\}$ orbit.

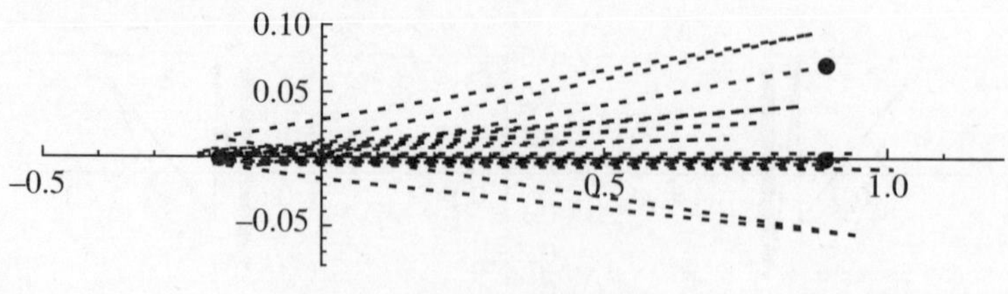

Fig. 6.3.3 An attracting 2-cycle

The next image superimposes the orbit of Fig. 6.3.3 on the fractal to show how it bounces between two regions of non-convergence.

Fig. 6.3.4 An attracting 2-cycle connecting two regions of non-convergence

Investigation 6.16 Explore the other 2-cycles corresponding to the equation, $z^3 - 2.55z^2 - 1 = 0$ and determine whether they are repelling or attracting.

Investigation 6.17 Identify and explore some 3-cycles corresponding to the equation, $z^3 - 2.55z^2 - 1 = 0$.

Non-convergence can also take the form of chaotic jumping around. Consider the simple equation which we considered earlier: $z^2 - 1 = 0$. The Newton–Raphson iteration for this is

$$z_{k+1} = \frac{1}{2}\left(z_k + \frac{1}{z_k}\right)$$

If we take z_1 on the imaginary axis then all the subsequent points of the orbit are also on the imaginary axis. Let $z_k = iy_k$. Then,

$$y_{k+1} = \frac{1}{2}\left(y_k - \frac{1}{y_k}\right)$$

Imagine a large value of y_1. Then y_2 will be approximately half of y_1. This successive halving will bring the orbit near $y_k = 1$ from where it will jump to near 0 and then away again to a large magnitude. We illustrate this below for the orbit starting with $y_1 = 10$.

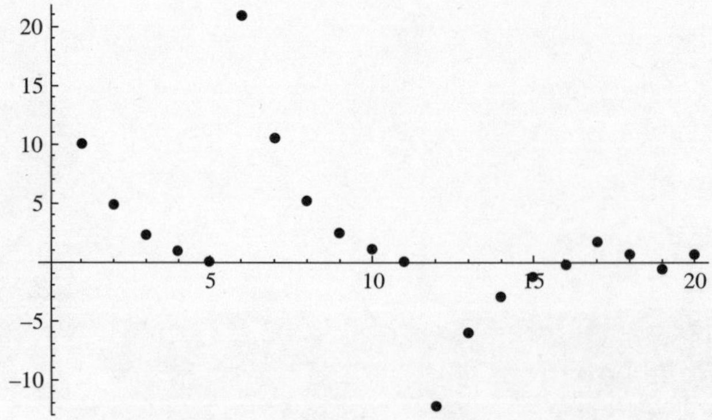

Fig. 6.3.5 An orbit with unpredictable jumps

The jumps away from 0 can be arbitrarily large. Let us view the same orbit over many more steps.

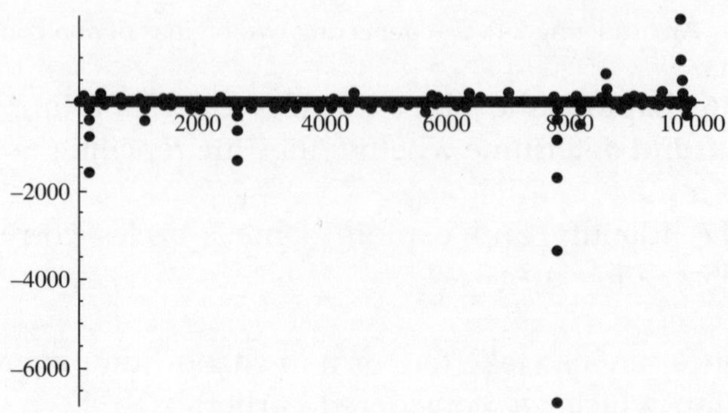

Fig. 6.3.6 A long term view of the orbit of Fig. 6.3.5

Such orbits are called 'chaotic'.

Investigation 6.18 Compute the first 10,000 members of the orbit starting with $y_1 = 10$ and plot a histogram of the values.

The next investigation will introduce you to a different kind of chaotic orbit—one which is initially orderly and suddenly switches to random looking jumps.

Investigation 6.19 Verify that $\pm i/\sqrt{3}$ is a 2-cycle for the Newton–Raphson iteration of $z^2 - 1 = 0$. Plot 100 iterations. Is it attracting or repelling?

One may wonder if an orbit can steadily increase in magnitude. This is impossible for the Newton–Raphson iteration of a polynomial equation. Suppose

$f(z) = z^n + a_{n-1}z^{n-1} + \cdots + a_0$. Then

$$N(z) = z - \frac{f(z)}{f'(z)} = z - \frac{z^n + \cdots}{nz^{n-1} + \cdots}$$

For large $|z|$ we have

$$N(z) \approx z - \frac{z^n}{nz^{n-1}} = (1 - 1/n)z \implies |N(z)| < |z|$$

Let us list the kinds of orbits we have discovered for the Newton–Raphson method applied to polynomial equations.

(i) Cyclic orbits which repeatedly go over the same points. (Including the possibility of staying at one point, which happens if we are lucky enough to start at a solution.)

(ii) Orbits which converge to cyclic orbits. (Including the possibility of converging to an orbit with just one point, that is, a solution.)

(iii) Chaotic orbits.

If we slightly perturb the starting point of an orbit, the orbit may change in one of two ways. The first is that the perturbed orbit stays close to the original one. This happens if the original point was in a basin of attraction for a cyclic orbit. The other is that the orbit changes drastically from the original. This happens if the original point was on the boundary between basins of attraction. Points of the second kind constitute the 'Julia set' for the iteration.

For example, the Julia set for the Newton–Raphson iteration of $z^2 - 1 = 0$ is the imaginary axis.

`Mathematica` has a built-in command for visualising the Julia set of a given iteration. We use it below to depict the Julia set for $z^3 - 1 = 0$:

```
JuliaSetPlot[z-(z^3-1)/(3*z^2), z, MaxIterations->100,
 ColorFunction->ColorData[{"GrayTones","Reverse"}]]
```

To depict the Julia set, we consider how many iterations it takes an orbit to get close to a solution, within some cut off (set at 100 in the code above). Points near the boundary would take longer to do so, and are accordingly given darker colours. Thus the Julia set is in the darkest coloured portion. Here are the results for some of the fractal patterns we have already considered.

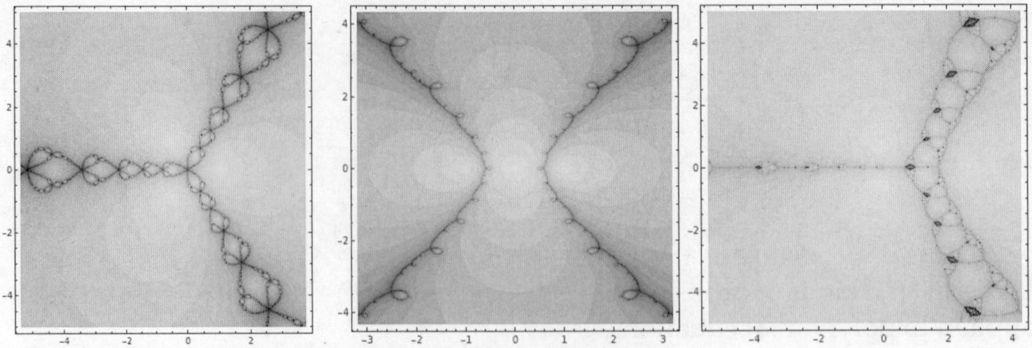

Fig. 6.3.7 Julia sets for the Newton–Raphson iterations for $z^3 - 1 = 0$, $z^3 - z = 0$ and $z^3 - 2.55z^2 - 1 = 0$

6.4 POLYNOMIAL ITERATION

The Newton–Raphson iteration has the form $z_{n+1} = \Phi(z_n)$ where Φ is a certain rational function (that is, a ratio of polynomials), and the goal is to find 'fixed points' satisfying $\Phi(z) = z$. We could attempt a similar exploration of orbits for any choice of function Φ.

The first question is whether a particular fixed point a is attracting or repelling. We observe the following:

$$z_{n+1} - a = \Phi(z_n) - \Phi(a) \approx \Phi'(a)(z_n - a)$$

Hence $|\Phi'(a)| < 1$ would make the gap between a and z_n shrink further, once z_n was close enough to a. Similarly, $|\Phi'(a)| > 1$ would prevent the gap from shrinking beyond a point. Therefore,

- A fixed point a is attracting if $|\Phi'(a)| < 1$.
- A fixed point a is repelling if $|\Phi'(a)| > 1$.

Investigation 6.20 Show that all fixed points of a Newton–Raphson iteration are attracting.

We introduce the following notation for the points of the orbit of z:

$$\Phi^n(z) = \underbrace{\Phi(\Phi(\cdots(\Phi}_{n \text{ times}}(z))\cdots))$$

Now suppose a is one of the points of an n-cycle. Then $\Phi^n(a) = a$ and a is a fixed point of Φ^n. Moreover, $(\Phi^n)'$ has the same value at every point of the n-cycle, since

$$(\Phi^n)'(a) = \Phi'(\Phi^{n-1}(a)) \cdot \Phi'(\Phi^{n-2}(a)) \cdots \Phi'(a)$$

and the right hand side is the product of the derivatives at all the points of the cycle and does not depend on the choice of a. This leads to the following criteria.

- An n-cycle is attracting if $|(\Phi^n)'(a)| < 1$ for any point a in the cycle.
- An n-cycle is repelling if $|(\Phi^n)'(a)| > 1$ for any point a in the cycle.

The following result due to Pierre Fatou will be very useful in explaining some of the phenomena we are encountering in our explorations.

Theorem 6.1 If Φ is a rational function (a ratio of polynomials) then for every attracting cycle of Φ there is a critical point of Φ whose orbit converges to that cycle.

Investigation 6.21 Show that the Newton–Raphson iteration of a cubic polynomial has at most one attracting cycle, apart from the fixed points.

In the remainder of this chapter, we shall consider Φ to be a polynomial whose highest degree coefficient is 1.

If $\Phi(z) = z + a$ is linear, then $\Phi(z_n) = z_0 + na$ moves linearly away from z_0 and there are no interesting phenomena.

If $\Phi(z) = z^2 + 2az + b$ is quadratic, we first complete squares:

$$\Phi(z) = (z^2 + 2az + a^2) + (b - a^2) = (z + a)^2 + c$$

Thus, by a shift of the origin, we can assume that there is no degree one term and $\Phi(z)$ equals $z^2 + c$.

Let us first consider the simplest situation: $\Phi(z) = z^2$. Then $\Phi^n(z_0) = z_0^{2^n}$. If $|z_0| < 1$ then the orbit converges to 0. If $|z_0| > 1$ the orbit moves farther and further away from 0 and we say that it 'converges to infinity'. If $|z_0| = 1$ the orbit stays on the unit circle.

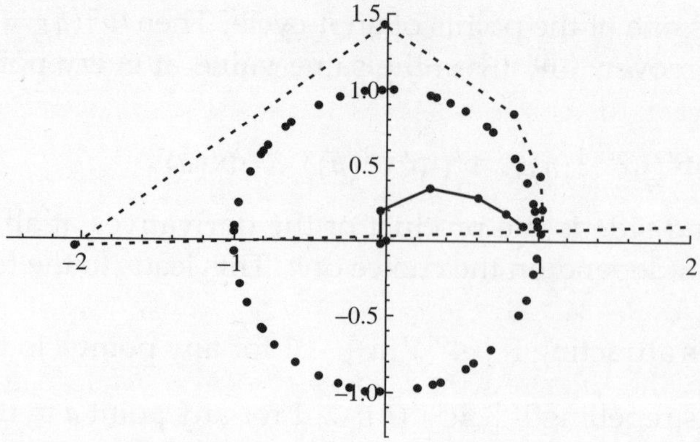

Fig. 6.4.1 Orbits of $\Phi(z) = z^2$ showing three kinds of behaviour depending on whether the starting point is inside, on, or outside the unit circle

We view the complex plane as consisting of two basins of attraction, one for zero and one for infinity, separated by the unit circle. The unit circle is the Julia set for this iteration.

Consider a polynomial $\Phi(z) = z^n + a_{n-1}z^{n-1} + \cdots + a_0$ of degree at least 2. Then the dominant term in $\Phi(z)/z$ is z^{n-1}, and so $|\Phi(z)/z| > 2$ if $|z|$ is large enough. Thus, there is a magnitude M such that $|z_0| > M$ guarantees that the orbit defined by $\Phi^n(z_0)$ converges to infinity. This shows that in polynomial iteration, there is always a basin of attraction for infinity. The complement of the basin of attraction for infinity is called the 'filled Julia set' and it lies completely inside some circle. Further, the Julia set is the boundary separating the filled Julia set and the basin of attraction for infinity.

`Mathematica` can provide both the Julia set and the filled Julia set. We give below the code for generating these sets for $\Phi(z) = z^2 - 1$.

```
{JuliaSetPlot[z^2-1, z, ColorFunction->None, PlotStyle->Black],
 JuliaSetPlot[z^2-1, z, ColorFunction->ColorData[{"GrayTones","Reverse"}]]}
```

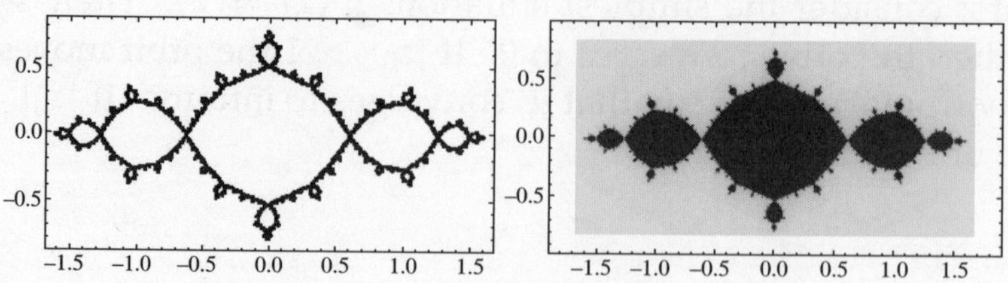

Fig. 6.4.2 The Julia set and filled Julia set for $\Phi(z) = z^2 - 1$

In the remainder of this section we shall concentrate on the filled Julia set and how it changes with c. We know that for $c = 0$ it is the solid unit disc centred at the origin. We first track the changes as c takes on the positive values 0.1, 0.2, 0.3 and 0.8.

```
Table[JuliaSetPlot[z^2+c,z,PlotRange->{{-1.3,1.3},{-1.3,1.3}},
 ColorFunction->ColorData[{"GrayTones","Reverse"}],
 MaxIterations->100],{c,{0.1,0.2,0.3,0.8}}]
```

As we increase c, the Julia set (the boundary of the black region) becomes more crinkly and fractal. At some point it loses its contiguity and its interior merges with the basin of attraction for infinity.

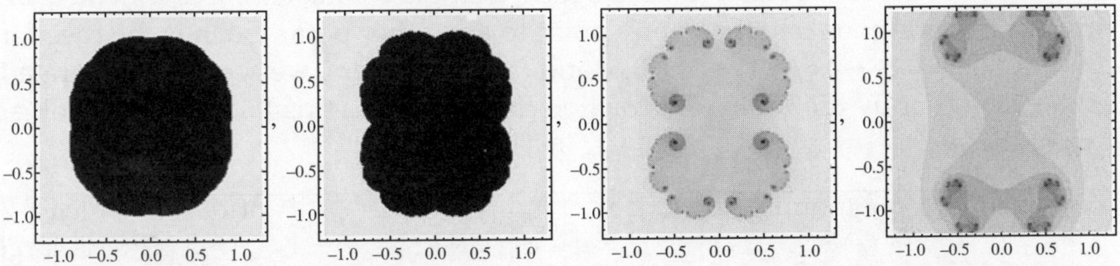

Fig. 6.4.3 Filled Julia sets of $\Phi(z) = z^2 + c$ with $c = 0.1, 0.2, 0.3$ and 0.8

Investigation 6.22 Consider various values of c between 0.2 and 0.3 to try and pin down when the Julia set stops separating the plane into two parts. (Resolution can be improved by increasing the `MaxIterations` setting.)

Next, we let c take the negative values -0.3, -0.6, -1 and -1.3. Again, the boundary becomes more fractal, but in this case it does not lose its contiguity and continues to separate the interior from the basin of attraction for infinity. The last two of these are called the 'basilica' and 'airplane', respectively.

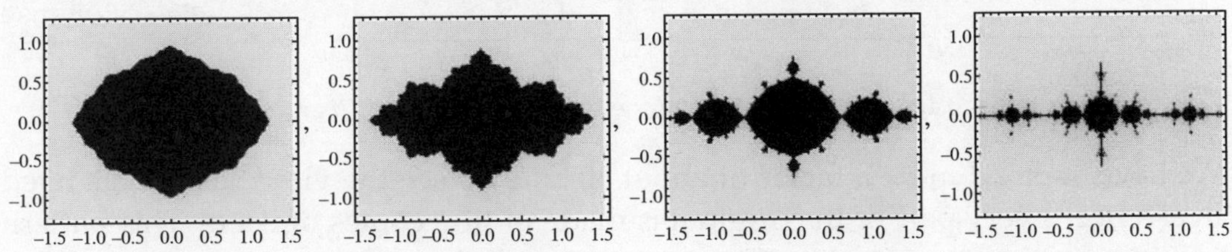

Fig. 6.4.4 Filled Julia sets of $\Phi(z) = z^2 + c$ with $c = -0.3, -0.6, -1$ and -1.3

Investigation 6.23 Consider further negative values of c like -1.5, -2, and so on. Does the Julia set eventually shatter?

Finally, we consider the imaginary values $0.6i$, $0.7i$, $0.85i$ and i. We see that the (filled) Julia set disintegrates somewhere between $c = 0.6i$ and $c = 0.7i$, but then pulls together again by $c = i$.

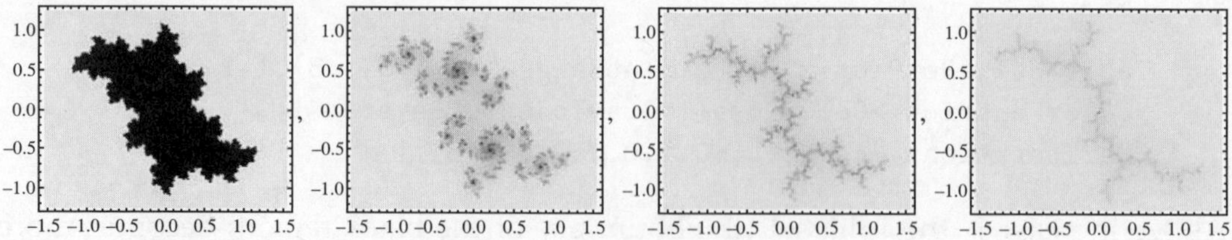

Fig. 6.4.5 Filled Julia sets of $\Phi(z) = z^2 + c$ with $c = 0.6i, 0.7i, 0.85i$ and i

We see that some of the filled Julia sets form a single connected piece in which we can plot a continuous path from any point to any other point. Such is the case for $c = 0.1, 0.2, -0.3, -0.6, -1, -1.3, 0.6i$ and i. In the others, we see parts separated from each other by the basin of attraction for infinity. Let us take a closer look at two of these.

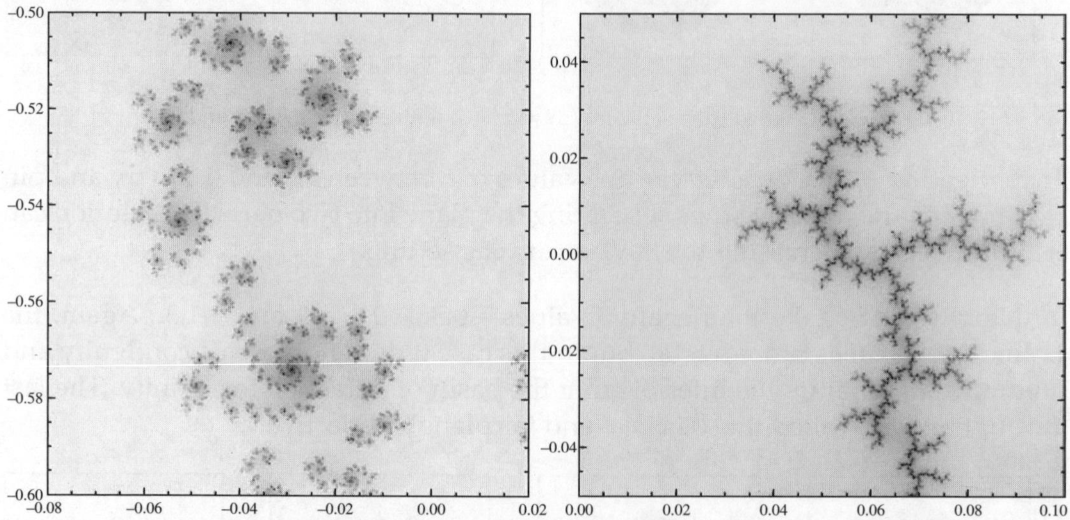

Fig. 6.4.6 Close-up views of the Julia sets with $c = 0.7i$ and $0.85i$

We have zoomed in by a factor of about 30 and we see the view has not changed in essence. Every part of the original is made of tiny copies of itself! This pattern persists if we zoom in further. The consequence is that no two points of these Julia sets can be joined by a continuous path which lies within the set.

A remarkable fact is that every quadratic iteration falls within one of these extremes—either every pair of points in the filled Julia set can be joined by a continuous path within the set, or none can! This is called *the fundamental dichotomy*. Julia sets of the second kind are called 'Cantor sets' or 'Cantor dust'.

Investigation 6.24 Compute the orbit of the origin in some of the above cases. Is there a correlation between the nature of the orbit and the connectivity of the Julia set?

We close this section with images of two famous Julia sets. The first is the 'Douady Rabbit', corresponding to $c = -0.13 + 0.75i$. The second is the 'Seahorse', with $c = -0.745 - 0.113i$. The close-up reveals the reason for its name.

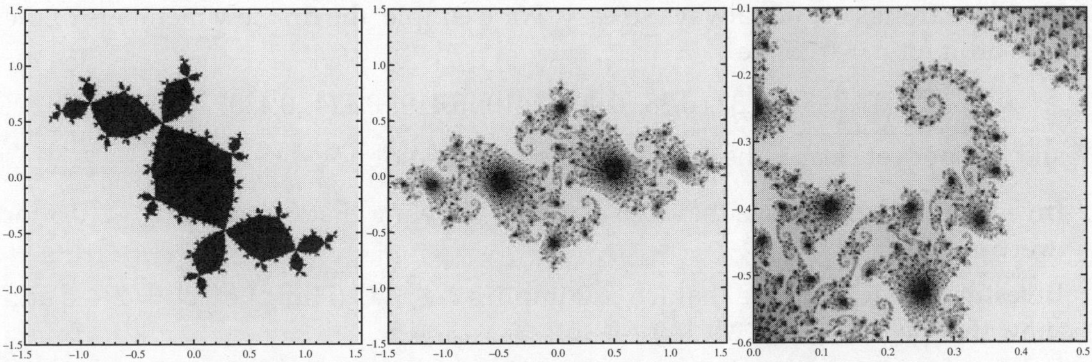

Fig. 6.4.7 The Douady Rabbit and two views of the Seahorse

6.5 THE MANDELBROT SET

In the final section of the chapter, we shall explore how Julia sets corresponding to the quadratic iteration $\Phi_c(z) = z^2 + c$ change with c. We shall focus on the fundamental dichotomy and seek the c for which the filled Julia set is connected, that is, any two of its points can be joined by a continuous path. Fortunately, Gaston Julia established a simple test in 1918.

Theorem 6.2 The fundamental dichotomy for the quadratic iteration $\Phi_c(z) = z^2 + c$ can be resolved as follows:

(i) If the orbit of zero is bounded then the filled Julia set is connected.

(ii) If the orbit of zero converges to infinity then the filled Julia set is totally disconnected.

The *Mandelbrot set* $\mathcal{M}$ is the collection of all c such that the filled Julia set is connected. For example, consider the following:

$$c = 0 : 0 \to \Phi_c(0) = 0^2 + 0 = 0 \to \Phi_c(0) = 0 \to 0 \to 0 \cdots \implies 0 \in \mathcal{M},$$
$$c = 1 : 0 \to \Phi_c(0) = 0^2 + 1 = 1 \to \Phi_c(1) = 2 \to 5 \to 26 \cdots \implies 1 \notin \mathcal{M},$$
$$c = -1 : 0 \to \Phi_c(0) = 0^2 - 1 = -1 \to \Phi_c(-1) = 0 \to -1 \to 0 \cdots \implies -1 \in \mathcal{M},$$
$$c = -2 : 0 \to \Phi_c(0) = 0^2 - 2 = -2 \to \Phi_c(-2) = 2 \to 2 \to 2 \cdots \implies -2 \in \mathcal{M},$$
$$c = i : 0 \to \Phi_c(0) = 0^2 + i = i \to \Phi_c(i) = -1 + i \to -i \to -1 + i \cdots \implies i \in \mathcal{M}.$$

In the specific cases so far, the orbit of zero either settled into a cycle or very quickly blew up. Things are not always so easy. For example, the first few members of the zero orbit for $c = 0.25$ are

$$0,\ 0.25\ ,0.31,\ 0.34,\ 0.37,\ 0.38,\ 0.40,\ 0.410,\ 0.418,\ 0.424,\ 0.430,\ 0.435,\ 0.439,\ \ldots$$

and it is not yet clear if the orbit will stay bounded or not.

Investigation 6.25 Prove that $0.25 \in \mathcal{M}$ by showing that $0 < \Phi^n_{0.25}(0) < 0.5$ for every n.

Investigation 6.26 Prove that $|c| > 2$ implies $c \notin \mathcal{M}$. (Hint: Let $|c| = 2 + \epsilon$ and show that $|\Phi^n_c(0)| \geq 2 + 2^{n-1}\epsilon$.)

Hence $\mathcal{M}$ is contained in the $|z| \leq 2$ disc. It does not completely fill the disc, since $1 \notin \mathcal{M}$.

The next investigation shows that if the zero orbit ever leaves the $|z| \leq 2$ disc, it will then converge to infinity. This provides a practical (but not perfect) test for membership in $\mathcal{M}$.

Investigation 6.27 Let $|c| \leq 2$. Show that if $|\Phi^n_c(0)| = 2 + \epsilon > 2$ for some n then $|\Phi^{n+k}_c(0)| \geq 2 + 4^k\epsilon$ for every $k = 1, 2, 3, \ldots$.

Investigation 6.28 Prove that $\mathcal{M}$ is symmetric about the x-axis. We have $c \in \mathcal{M}$ if and only if $\bar{c} \in \mathcal{M}$.

We would like to proceed methodically to discover points of $\mathcal{M}$. Fatou's theorem about attracting cycles will prove helpful here. First we note that the only critical point of Φ_c is $z = 0$. Hence if Φ_c has an attracting cycle, the orbit of 0 must converge to it, and then the orbit will be bounded and we shall have $c \in \mathcal{M}$.

Let us first look for c such that Φ_c has an attracting fixed point w. Then we have

$$\Phi_c(w) = w \iff w^2 + c = w \quad \text{and} \quad |\Phi'_c(w)| < 1 \iff |w| < 1/2$$

Hence c takes the values $w - w^2$ where w varies over the open disc with radius $1/2$ and centre at origin. The points w can be expressed as $w = re^{i\theta}$ with $0 \le r < 1/2$, giving

$$c = re^{i\theta} - r^2 e^{2i\theta} = (r\cos\theta - r^2\cos 2\theta) + i(r\sin\theta - r^2\sin 2\theta).$$

We ask `Mathematica` for the plot:

```
fixed=ParametricPlot[{r*Cos[t]-r^2*Cos[2*t], r*Sin[t]-r^2*Sin[2*t]},
 {t,0,2*Pi}, {r,0,0.5}, PlotStyle->GrayLevel[0.5], BoundaryStyle->None];
fixededge=ParametricPlot[{Cos[t]/2-Cos[2*t]/4, Sin[t]/2-Sin[2*t]/4},
 {t,0,2*Pi}, PlotStyle->Black];
Show[{fixed,fixededge}, PlotRange->{{-1,.4},{-.7,.7}}, Axes->True,
 Frame->False, AspectRatio->1]
```

The result is a shape known as a 'cardioid'.

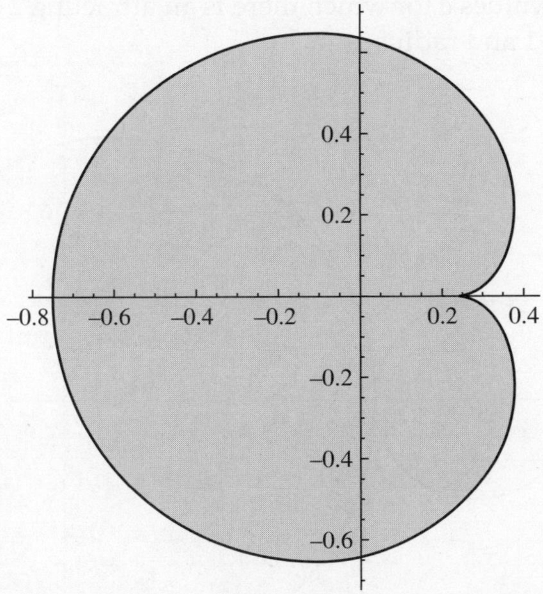

Fig. 6.5.1 The points c in the interior of this cardioid correspond to Φ_c having an attracting fixed point. The cardioid intersects the x-axis at $1/4$ and $-3/4$.

Next, we look for attracting 2-cycles. Suppose $\{w, \Phi_c(w)\}$ is a 2-cycle. That is, $\Phi_c^2(w) = w$. This gives a degree 4 polynomial equation:

$$(w^2 + c)^2 + c = w \quad \text{or} \quad w^4 + 2cw^2 - w + c^2 + c = 0$$

Since the fixed points are also solutions of this equation, $w^2 - w + c$ must be a factor. Dividing by it gives

$$w^4 + 2cw^2 - w + c^2 + c = (w^2 - w + c)(w^2 + w + c + 1)$$

So there should be a single 2-cycle $\{w, \Phi_c(w)\}$, satisfying

$$w^2 + w + c + 1 = 0 \tag{6.8}$$

The condition for the cycle to be attracting, is $|(\Phi_c^2)'(w)| < 1$. That is,

$$|4w(w^2 + c)| < 1 \quad \text{or} \quad |w(w^2 + c)| < 1/4$$

Substituting Eq. (6.8) gives

$$|w(w + 1)| < 1/4 \quad \text{or} \quad |w^2 + w| < 1/4$$

We can eliminate w by substituting Eq. (6.8) again:

$$|c + 1| < 1/4$$

Thus the parameter values c for which there is an attracting 2-cycle, form an open disc with centre at -1 and radius $1/4$.

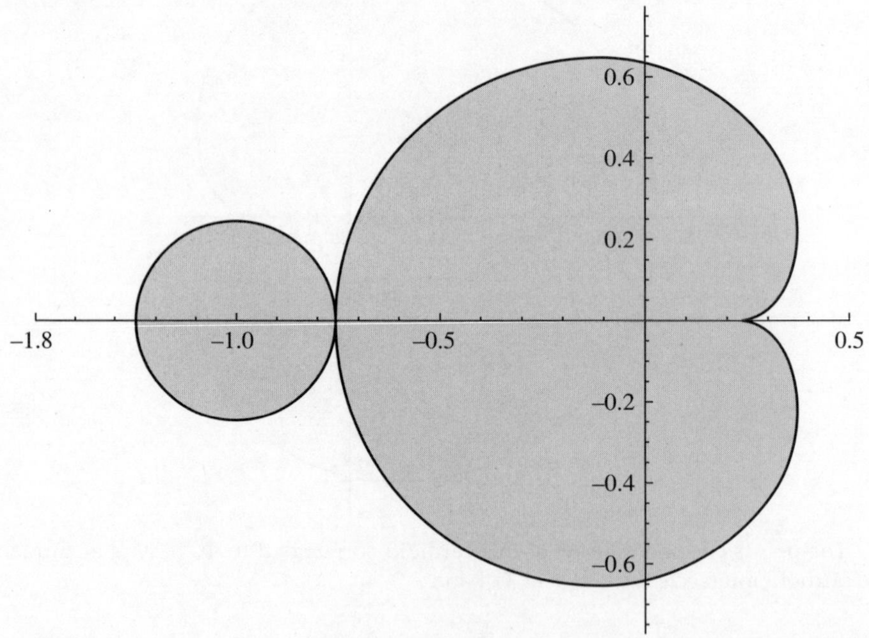

Fig. 6.5.2 A combined plot of the regions of attracting fixed points and 2-cycles

Longer attracting cycles are harder to work out theoretically, and we won't attempt it. However, we should expect them to create further little blobs that would be tacked on to the main shape we have already discovered. Moreover there would be points due to repelling cycles as well as orbits which are bounded but not cyclic. As we shall shortly see, these introduce tremendous intricacy to the structure of the Mandelbrot set.

We close our theoretical work with an investigation of the real numbers which are in the Mandelbrot set. We have already proved that $[-5/4, 1/4] \subset \mathcal{M}$. Let us illustrate our calculations with some orbit diagrams. First, we see how $c \in (-0.75, 0.25)$ leads to orbits that converge to a fixed point. Further, the $c = 0.26$ orbit is seen to diverge, to illustrate that points $c > 0.25$ are not in $\mathcal{M}$.

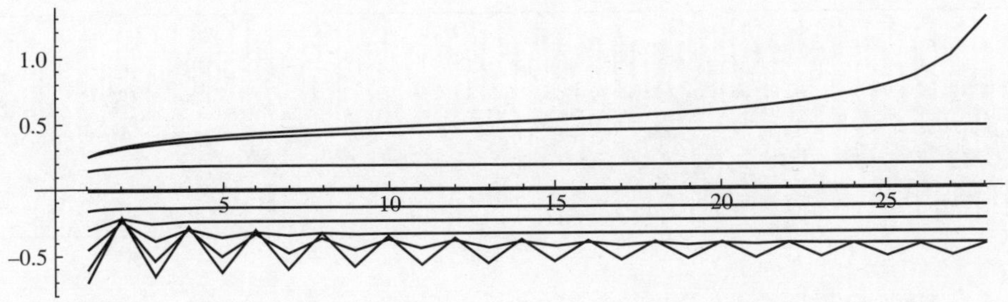

Fig. 6.5.3 Orbits of zero, corresponding to $c \in (-0.75, 0.25)$, converge. The topmost orbit, corresponding to $c = 0.26$, goes to infinity.

Investigation 6.29 Show that if $c \in \mathbb{R}$ then $\Phi_c^{n+1}(0) - \Phi_c^n(0) \geq (c - 1/4)$. Hence $c > 1/4$ would imply that the zero orbit converges to infinity.

Next, we consider two orbits with $c \in (-1.25, -0.75)$ and see that they converge to 2-cycles, as already established.

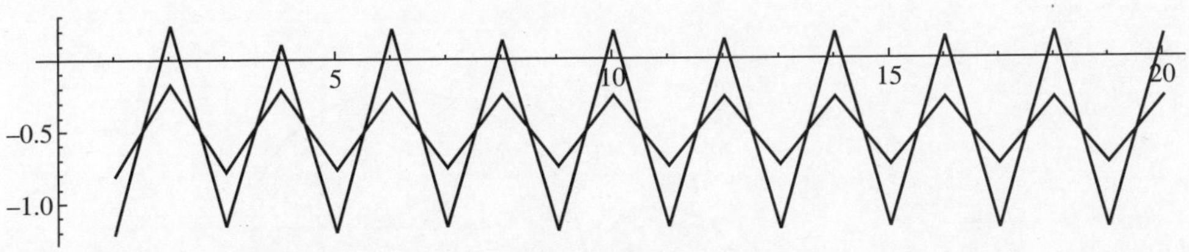

Fig. 6.5.4 Orbits of zero, corresponding to $c \in (-1.25, -0.75)$, converge to 2-cycles

We have not yet established anything about zero orbits with $c \in (-2, -1.5)$. Let us plot two examples.

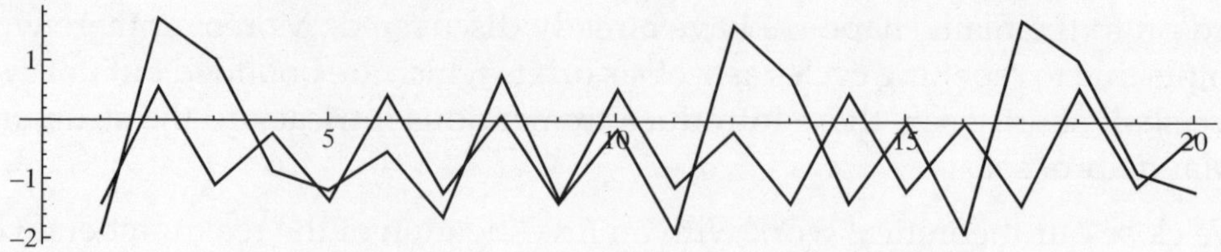

Fig. 6.5.5 These orbits with $c \in (-2, -1.5)$ stay bounded although their behaviour is chaotic

Let's take a longer look at an orbit from the chaotic range, the one with $c = -1.7$.

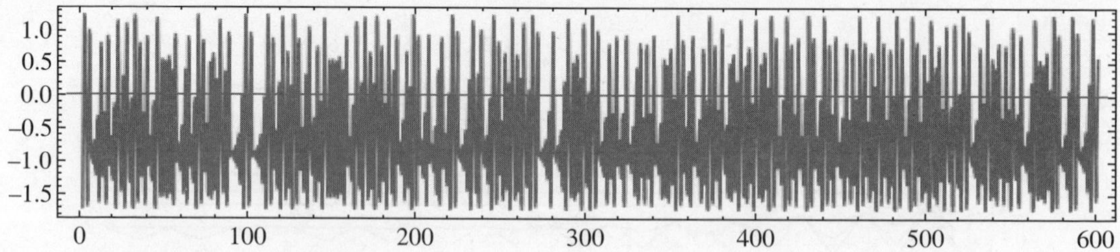

Fig. 6.5.6 The first 600 points of the zero orbit corresponding to $c = -1.7$

We see that the later peaks do not seem to cross the initial ones. This suggests the following approach for proving boundedness.

Investigation 6.30 Show that $-2 \leq c \leq 0$ implies $|\Phi_c^n(0)| \leq |c|$ for every n.

Thus, we have so far identified a cardioid 'body', a circular 'head', and a spike on top.

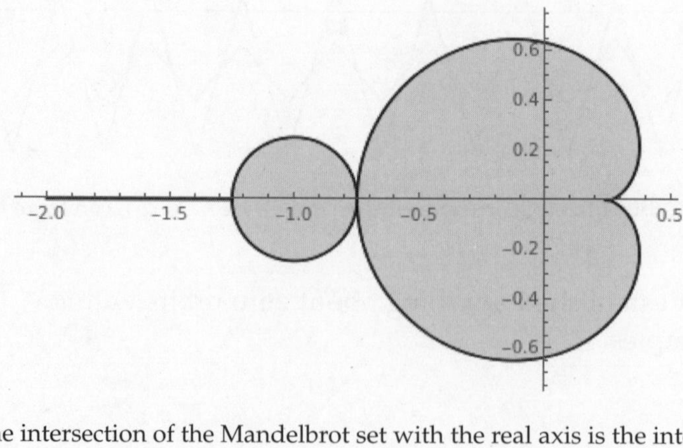

Fig. 6.5.7 The intersection of the Mandelbrot set with the real axis is the interval $[-2, 0.25]$

Now we ask `Mathematica` to show us the Mandelbrot set in full, by typing

```
MandelbrotSetPlot[ColorFunction->ColorData[{"GrayTones","Reverse"}]]
```

We can also ask `Mathematica` to show us the part of the Mandelbrot set lying in a rectangle with given opposite vertices. For example, if the opposite vertices are -1.5 and i, we type (skipping the options for colour chocies):

```
MandelbrotSetPlot[{-1.5,I}]
```

The full Mandelbrot set is shown in Fig. 6.5.8 with the already known parts superimposed on it. The known member i is indicated by a + symbol. Note how far it is from the main body, although it appears to be connected to it by a fibrous structure.

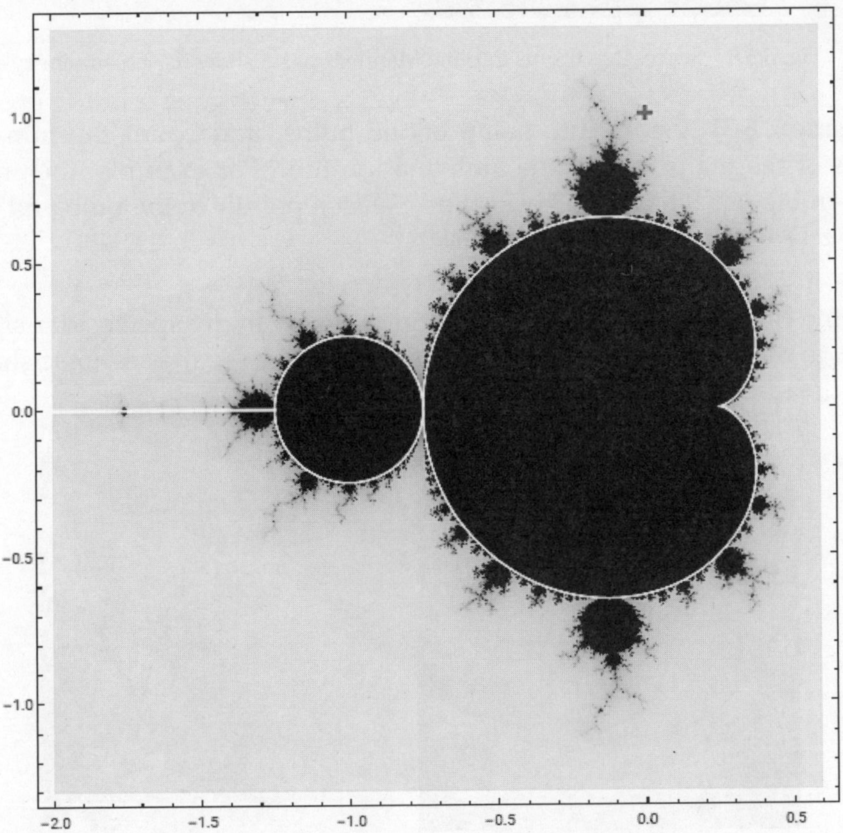

Fig. 6.5.8 A detailed look at the Mandelbrot set

The very first thing we notice is the scattering of bulbs of various sizes on the surface of the main body and head, and even on each other, with each bulb looking quite similar to the full Mandelbrot set. Moreover, tiny copies of the Mandelbrot set seem to be present even along the fibres extending out from the main body.

Fig. 6.5.9 Successive zooms into the Mandelbrot set show its self-similarity

Investigation 6.31 Zoom into some of the bulbs, and count the number n of branches of the main fork of the 'antenna' on top. For example, the antenna of the bulb in Fig. 6.5.9 has $n = 3$ branches. Select a point c in the bulb and compute whether $z^2 + c$ has an attracting n-cycle.

Extreme close-ups of the boundary region reveal many spectacular sights. For example, in the space between the body and head, we see the seahorse shapes that we had already observed among the Julia sets.

Fig. 6.5.10 Views of the 'Seahorse Valley'

Zooming in further requires so much more computation that it is better done with dedicated programs.

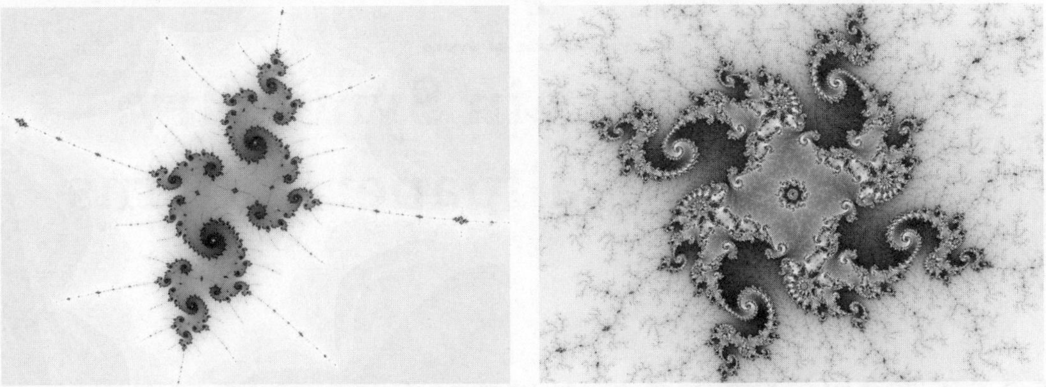

Fig. 6.5.11 Extreme zooms into the Mandelbrot set, produced by David Eck's `Mandelbrot Viewer` [11].

FURTHER EXPLORATIONS

The 1991 articles by Straffin [69] and Strang [70] were among the first to make this material accessible to school and college students. They contain valuable insights and further explorations, especially of the emergence of chaos. McClure [33] is a more recent exposition which uses `Mathematica` and implements a method for systematically finding all the roots of a complex polynomial. The 1984 article by Peitgen, Saupe and v Haeseler [58] gives examples of using computer graphics to discover mathematical facts. Yau and Ben-Israel [90] show how to visualise the Newton–Raphson method over complex numbers, and also describe the related Halley method.

John Whitehouse's website [84] has many images, and a computer program, for the fractals arising from the Newton–Raphson method. David Eck's website [11] does the same for the Mandelbrot set. Many interesting videos are available online. We particularly recommend the *Mandelbrot Set Explained Series* [32] on the Mathematicians' Guild channel.

For a general exposition of fractals and their presence in nature, the classic text is Mandelbrot's own *The Fractal Geometry of Nature* [31], which is an expanded and updated version of his earlier book [30].

7

Experiments in Symmetry: Frieze and Wallpaper Patterns

7.1 The Ubiquitous Nature of Symmetry

Symmetry is all around us, to be experienced in nature and in life. From the designs on the clothes we wear, to railings and grills of staircases and windows, the pavements and tiles that line our roads to the in-depth three dimensional view of a virus or a computer constructed fractal, we cannot escape the fact that symmetry is ever present. Ancient monuments, irrespective of eras, countries and rulers, prove to be excellent examples of how symmetry plays a role in determining both structure and decorations. Indeed, nature abounds with examples of why certain symmetries are not only beautiful but also are the most efficient design for the required function. It is not a coincidence that the 'genes' of nature are mathematical.

Symmetry has become synonymous with beauty. One would be remiss in discussing symmetry without mentioning the work of the artist M C Escher. You can find out more about Escher's life at [44] and the gallery of his work is available at [35]. Escher was very interested during a certain phase, in being able to draw repeated motifs in such a manner as to fill the drawing sheet or as it is termed, 'filling or tiling the plane'. He visited the Alhambra Palace several times and the decorations there [1] served as his inspiration. The last section of this chapter deals with a connected concept, namely, that of wallpaper patterns.

In this chapter we will look at symmetry rather closely, using the dual lens of mathematics and technology. The main focus is going to be on studying 'Frieze or Strip Patterns and Wallpaper Patterns'. Starting from how these are defined, we consider the different types of strip patterns and wallpaper patterns. We will use the mathematics of symmetry for this classification. Technology will play its part in visualising these patterns and in implementing algorithms to systematically identify the 'type' of a given pattern. We shall also see how this can play a part in creating patterns by using reverse engineering in a sense. But before we focus on frieze or wallpaper patterns, let us examine symmetry of an object from an intuitive point of view.

Intuitively, the symmetry of any object is an action that you can perform which will bring back the object occupying the exact same location as before and looking exactly like it did before the action was performed. In other words, it is an action which you do, that seems to leave the object untouched. The action also has to be such that it does not involve breaking the object concerned and piecing it back together. A more formal definition will be given later.

For example, let us consider an equilateral triangle and ask ourselves what its symmetries are. You can visualise a paper cut-out of an equilateral triangle and draw its outline in your notebook. Place the cut-out within the outline drawn. Using the above intuitive definition, a symmetry of the equilateral triangle is an action you can perform on the cut-out so that it comes back and occupies the same outline.

Let us use some simple technology to experiment with the equilateral triangle. Open a file in `GeoGebra`. Choose the `Regular Polygon` option from the fourth menu option on the top. As soon as you mark two points, you will be asked to specify the number of vertices, enter 3 and an equilateral triangle will be created. The `Move` option from the first menu option on the top, allows you to move the triangle and to resize it by dragging vertices. The vertices of the equilateral triangle should be labelled A, B and C in the counter-clockwise sense. This is usually the default labelling which will appear. You can also right click on the labels using the `Move` option to rename them if needed.

Now we need to find the centroid of the equilateral triangle. Choose the `Midpoint` option from the second menu option on the top and mark the mid-points of at least two of the three sides. Then use the `Segment` option from the third menu option on the top to join the vertices to the mid point of the opposite sides. Use the `Intersect` option from the second menu option on the top to choose two of the altitudes and mark their point of intersection. Rename it 'O'. Now use the `Algebra` view on the right to un-click the midpoints of the sides and the line segments to be left with just the equilateral triangle ABC with centroid O. You can also right click on the triangle and hide its label (which as default will be 'poly 1'). We are now in a position to perform rotations of the triangle about O and to ascertain which ones will be symmetries by our definition.

To enable rotations we choose `Slider` from the tenth menu option on the top and choose angle button and name the angle θ from the symbols on the right corner. The values of θ will range from 0 degree to 360 degree. From the ninth menu option on the top, select `Rotate about Point`. It will tell you to choose the object to rotate first, in our case, triangle ABC and then the point about which to rotate the object, namely, O. Once these two are chosen, it will ask for the angle, choose θ from the symbol menu in the right corner, and the counter-clockwise direction of rotation. Now if you move the slider for angle θ from 0 degree to 360 degree you will see a replica of the triangle (named poly 1′) rotating. When the two coincide you have a symmetry. This will happen only when the angle is 0 degree, 120 degree and 240 degree.

However we also need to track our symmetry and we do so by labelling the vertices of poly 1′ as A', B' and C' respectively. (This can be done by rotating poly 1′ by say 10 degree. Use the 'Point on Object' option to create 3 points for the vertices of poly 1′. See Fig. 7.1.1.) When we perform a 120 degree or 240 degree rotational symmetry, we check which vertex of poly 1′ is coinciding with those of poly 1. This gives us a way of writing down the effect of the rotational symmetries.

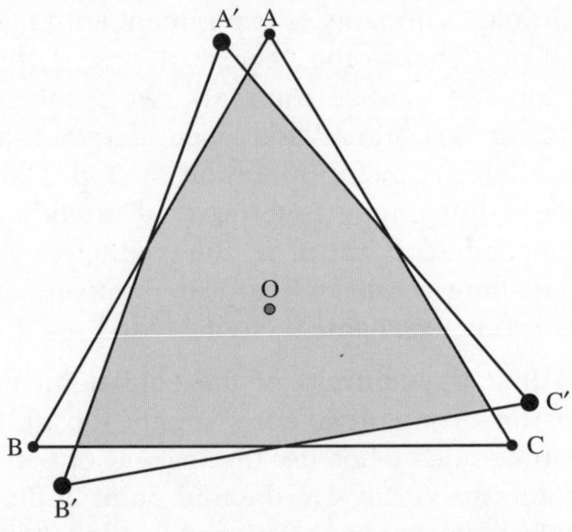

Fig. 7.1.1 Two copies of an equilateral triangle

Figure 7.1.2 shows the rotations of the equilateral triangle of degrees 0, 60, 120 and 240. It is obvious from the superimpositions of triangles ABC and $A'B'C'$ that except for the 60 degree rotation, the other three are symmetries.

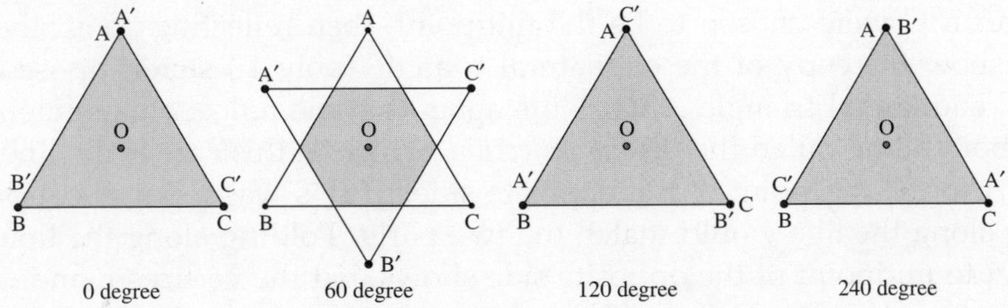

Fig. 7.1.2 Rotations of an equilateral triangle

Figure 7.1.2 also shows how we may keep track of the three rotational symmetries through the vertices of the equilateral triangle $A'B'C'$ as compared to those of the equilateral triangle ABC. So R_0 the zero degree or do nothing rotation, keeps all the vertices fixed. We can denote this as $R_0 = (A)(B)(C)$. The 120 degree rotation, R_{120} takes A to B (this can be seen as A' has moved to where B was), B to C and C to B. So we write $R_{120} = (A\ B\ C)$. In the same vein we get $R_{240} = (A\ C\ B)$ as it takes A to C, B to A and C to B.

Note that rotation is taking place in an anticlockwise direction about an axis perpendicular to the plane of the triangle about its centre which is the point of intersection of the lines joining vertices to mid-points of the opposite sides. This point about which the rotation is taking place is called the *'rotocenter'* and is the only point that remains fixed when a non-zero degree rotation is performed.

The other symmetry that is possible is a flip or a reflection. This can also be performed in `GeoGebra` by using the `Reflect about a Line` option from the ninth menu option on the top. Figure 7.1.3 below shows the reflection of the equilateral triangle ABC about a line g joining vertex A to a point G on the opposite side which is not the midpoint. The figure clearly shows that we do not have a symmetry.

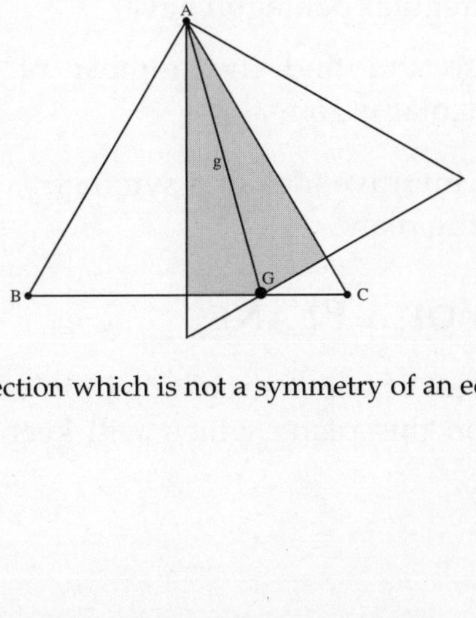

Fig. 7.1.3 A reflection which is not a symmetry of an equilateral triangle

However if G were chosen to be the midpoint, then reflecting about the line g, would show the copy of the equilateral triangle (poly 1′) superimposed on the original equilateral triangle ABC. Note again that the reflection symmetry takes place about a line called the *'line of reflection'* which in this case is the line joining vertex A to the midpoint of the opposite side. If this was a paper cut-out, then folding along the line would match the two parts. Folding along the line joining vertex A to midpoint of the opposite side shows that the vertices B and C would superimpose. We denote this reflection symmetry as $F_A = (A)(B\ C)$ as it fixes vertex A and interchanges B and C. There are only two other reflection symmetries possible for the equilateral triangle namely, $F_B = (B)(A\ C)$ about the line joining B to midpoint of AC and $F_C = (C)(A\ B)$ which is a reflection symmetry about the line joining C to the midpoint of the opposite side. Another observation is that all points on the line of reflection remain fixed when the reflection takes place.

Thus we see that there are six symmetries possible for an equilateral triangle X, namely $\text{Sym}(X) = \{R_0, R_{120}, R_{240}, F_A, F_B, F_C\}$, where $\text{Sym}(X)$ denotes the set of symmetries of the equilateral triangle X.

Investigation 7.1 Use `GeoGebra` to create a regular pentagon. Find its centre and rotate about it to find the rotational symmetries of a regular pentagon. Also draw suitable lines of reflection and ascertain that reflecting about them does give a symmetry of a regular pentagon.

(a) Write down the rotational symmetries of the regular pentagon labelled $ABCDE$ in the counter-clockwise sense by writing down its effect on the vertices as explained for an equilateral triangle. How many rotational symmetries does a regular pentagon have?

(b) Write down the reflection symmetries of the regular pentagon labelled $ABCDE$ in the counter-clockwise sense by writing down its effect on the vertices as explained for an equilateral triangle. How many reflection symmetries does a regular pentagon have?

(c) Can you generalise and find the number of rotational and reflection symmetries for a regular n-gon?

Now that we have an intuitive idea of a symmetry, we can apply it to finding the symmetries of the entire plane.

7.2 Symmetries of a Plane

Imagine the plane as a sheet of paper stretching infinitely and think of the actions that can be performed on this plane which will keep the sheet occupying the exact same position as before when considered as a whole. Again no tearing and re-constituting is allowed. You may be able to figure out that if we draw a line on the plane and reflect about the line it will be a reflection symmetry of the plane. It is akin to placing an infinitely stretched mirror along the line, perpendicular to the plane and noting that the entire plane is seen when considering the original with its reflection. Similarly, if we identify a point on the plane and rotate the entire plane by an angle of θ radians about an axis through the point, perpendicular to the plane, in the counterclockwise direction then we will again have a rotation symmetry of the plane. These two are familiar symmetries as we have encountered them while discussing symmetries of an equilateral triangle.

An equilateral triangle is a finite figure. We can draw a rectangle around it, whereas the plane is an infinite figure. The equilateral triangle did not have any symmetries other than rotations and reflections. However, there is one more symmetry we can think of for a plane which is just not possible for finite figures like the triangle. This is a symmetry called a translation symmetry. This occurs when we move the plane along a vector. The plane moves a distance which is equal to the magnitude of the vector and in the direction of the vector. Think of the plane moving along the same direction as an arrow and to a distance equal to the length of the arrow. Because the plane is infinite, it will still as a whole occupy the exact same position in space as it did before.

How do we keep track of planar symmetries? In the case of the equilateral triangle it was enough to track the symmetry through the changes to vertices. In the case of the plane we can imagine an **R** marked on the plane and we can track the symmetry by its effect on the **R**. Figure 7.2.1 shows the reflection of the plane about a line CD created using `GeoGebra`. Open a new file with the grid showing. Import a picture of **R**, resize and position it as required on the grid. Using the `Line` option draw a line CD and then use the `Reflect about Line` option to reflect the picture of the R about the line CD. If the plane is reflected about the line CD then the picture below shows the position of the original R and its position after the reflection of the plane.

Similarly we can visualise rotation symmetry of the plane by creating a point E on the grid using the `Point` option and then using the `Rotate around Point` option to rotate the figure R around the point E by the required angle. Figure 7.2.2 shows the rotation of the plane in the anticlockwise direction by a 120 degree angle about an axis through the point E that is perpendicular to the plane. The figure shows the original R and the effect after this rotation is performed.

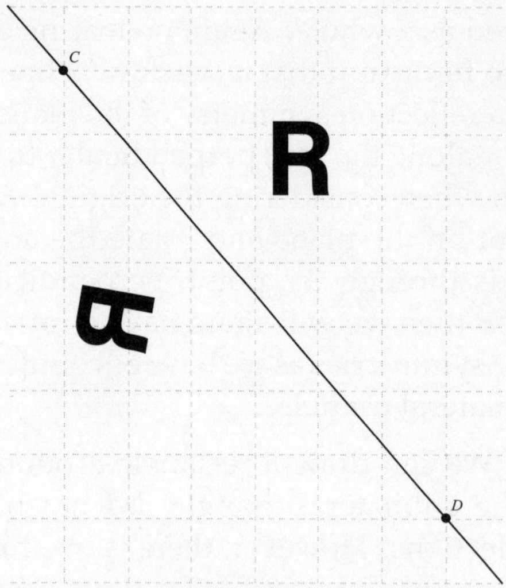

Fig. 7.2.1 Reflection of the plane

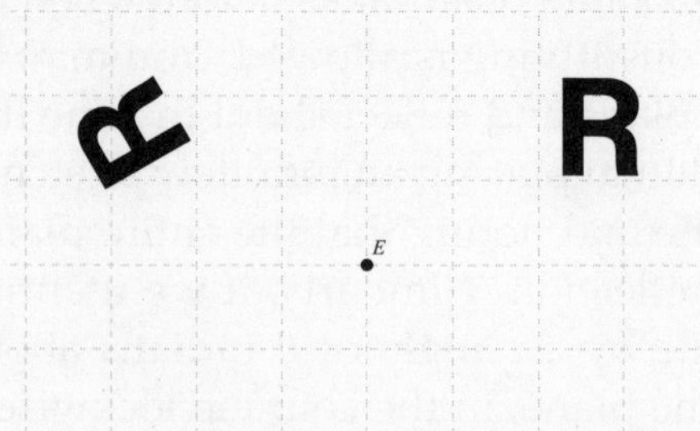

Fig. 7.2.2 Rotation of the plane

The translation symmetry we discussed can also be experienced using `GeoGebra`. This time we draw a vector u with initial and end points G and F respectively using the `Vector` option anywhere on the grid and then use the `Translate by Vector` option to translate the object R about the vector u. Figure 7.2.3 shows the translation of the plane along vector u by showing its effect on the letter R.

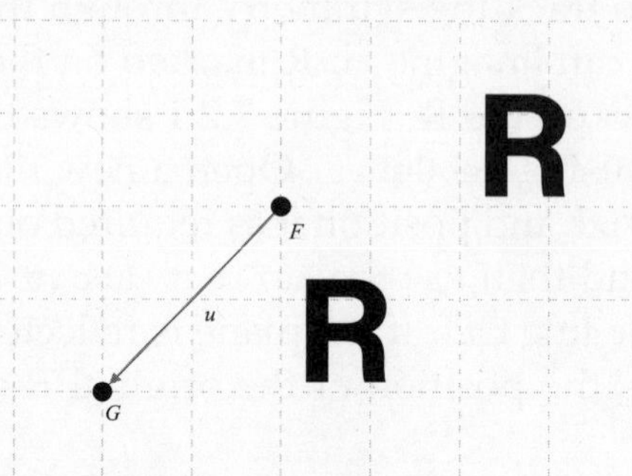

Fig. 7.2.3 Translation of the plane

We have now seen three types of symmetries for the plane—rotations, reflections and translations. We had also seen the two former symmetries when we considered the symmetries of an equilateral triangle. Indeed a little thought would tell us that for a finite planar figure like the equilateral triangle, it is not possible to have a translation symmetry. By finite figure, we mean figures around which we can draw a rectangle. Obviously due to the limits imposed by the finiteness of the figure, a translation would not leave the figure occupying the exact same location and looking the same as before. We now go back to our intuitive definition of the symmetry of an object. Recall that it is an action we can perform on the object leaving it occupying the exact same position and looking exactly the same as when we started. Also no breaking and reconstructing is allowed. Thus it is obvious that if a symmetry is followed by another symmetry the net result is again a symmetry.

For example, if we perform the 120 degree rotational symmetry of the equilateral triangle and then the reflection F_A, the result will be a symmetry again of the equilateral triangle. Let us denote this resultant symmetry by $F_A \circ R_{120}$ and by tracking the effect on the vertices let us find out what $F_A \circ R_{120}$ is. Recall that $F_A \circ R_{120} = (A)(B\ C) \circ (A\ B\ C)$. Thus we see that $F_A \circ R_{120}(A) = F_A(R_{120}(A)) = F_A(B) = C$. Thus $F_A \circ R_{120}$ takes the vertex A to C. It can be checked that it fixes B and takes vertex C back to A. Hence $F_A \circ R_{120} = (A\ C)(B) = F_B$.

In other words, if $\mathrm{Sym}(X)$ denotes the set of all the six symmetries of the equilateral triangle X, then for any $\sigma, \rho \in \mathrm{Sym}(X)$ we have that $\sigma \circ \rho \in \mathrm{Sym}(X)$. Thus there is *'closure'* under $\circ$, that is, when we apply one symmetry after the other, the result is again a symmetry. Indeed one can also check that R_0 has the property that $\sigma \circ R_0 = \sigma = R_0 \circ \sigma$ for any $\sigma \in \mathrm{Symm}(X)$. An element like R_0 which exhibits such a property is called an *'identity'* with respect to $\circ$. Further we see that for each $\sigma \in \mathrm{Sym}(X)$ we can find a corresponding $\rho \in \mathrm{Sym}(X)$ such that $\sigma \circ \rho = R_0 = \rho \circ \sigma$. For $\sigma = R_{120}$ we get $\rho = R_{240}$ and so on. We call ρ an *'inverse'* for σ. Another interesting observation we can make akin to that satisfied by addition on numbers is that if α, β and γ are all elements of $\mathrm{Sym}(X)$ then $(\alpha \circ \beta) \circ \gamma = \alpha \circ (\beta \circ \gamma)$. We say that $\circ$ is *associative*.

Now consider a non-empty set G with a good way $*$ of combining two elements of G to get an object of G again, like we have $\circ$ on Sym(X). Then we say that G is a *group* under $*$ if there is closure, associativity, existence of identity and inverses with respect to $*$. In other words, Sym(X) is a group under $\circ$. Another familiar example is the set of integers which is a group under addition with identity the number 0.

Investigation 7.2 We have seen that when X is an equilateral triangle, its group of symmetries is $G = \mathrm{Sym}(X) = \{R_0, R_{120}, R_{240}, F_A, F_B, F_C\}$. Create a 6×6 table where the rows and columns are indexed by the elements of G appearing in the same order. The entry in the table cell indexed by σ (row) and ρ (column) should be $\sigma \circ \rho$. Such a table is called a Cayley table for G. For example, the entry for the row indexed by F_A and column indexed by R_{120} will be F_B as $F_B = F_A \circ R_{120}$. (For more on Arthur Cayley see [42].)

(a) Draw and complete the Cayley table for the above G.

(b) Draw and complete the Cayley table for Sym(X) where X is a rectangle which is not a square.

(c) Can you think of a finite planar figure which has only 3 rotational symmetries? Is there a finite planar figure with exactly n rotational symmetries for any given natural number n? Is there a finite planar figure with infinitely many symmetries? For the first, draw a Cayley table.

We can actually show through more formal means that for any object X, if Sym(X) denotes the set of all its symmetries then it does form a group under $\circ$ with identity $I = R_0$, the 0 degree rotational symmetry or the do-nothing rotational symmetry. Now, let us continue with our exploration of symmetries of a plane. We know that if X is our plane then the collection of all its symmetries forms a group. Incidentally, this group of symmetries is called the group of rigid motions of the plane.

Via the investigation below, let us explore $\circ$ for symmetries of a plane.

Investigation 7.3 We have found three types of symmetries for a plane—rotation R about a point p, reflection F about a line l and translation T along a vector v. Use `GeoGebra` to explore the questions below. Rotations and translations below are assumed to be non-trivial, that is, different from the zero degree rotation or identity.

(a) Let R and R' be rotations of the plane about rotocenters p and p' respectively. Which of the three types of symmetry will $R \circ R'$ be? Is there a difference if $p = p'$ and when $p \neq p'$?

(b) Let F and F' be reflections about lines l and l' respectively. Which of the three types of symmetry will $F \circ F'$ be? Is there a difference if l and l' intersect versus when l and l' are parallel?

(c) If R is a rotation about a point p and F is a reflection about line l then which type of symmetry is $R \circ F$? Is there a difference when p lies on l and when it does not?

(d) If T is a translation about the vector v and R is a rotation about the point p then what can you say about $T \circ R$?

If you have worked through the above investigation you will notice that we have not talked about $T \circ F$ or $F \circ T$ where F is a reflection and T is a translation. This was deliberate because while we know that we have a symmetry in each of these two cases, it is not one of the three types that we saw earlier. This is a new symmetry type and is called a *glide reflection*. Let us explore the effect of a reflection

F about a line *l* followed by a translation *T* along a vector *v*, on the letter **R** using GeoGebra. Figure 7.2.4 shows three positions for **R**—a start position labelled **s**, an intermediate position labelled **i** after a reflection about line *l* followed by a finish position labelled **f** after the reflection is followed by a translation along a vector *v*.

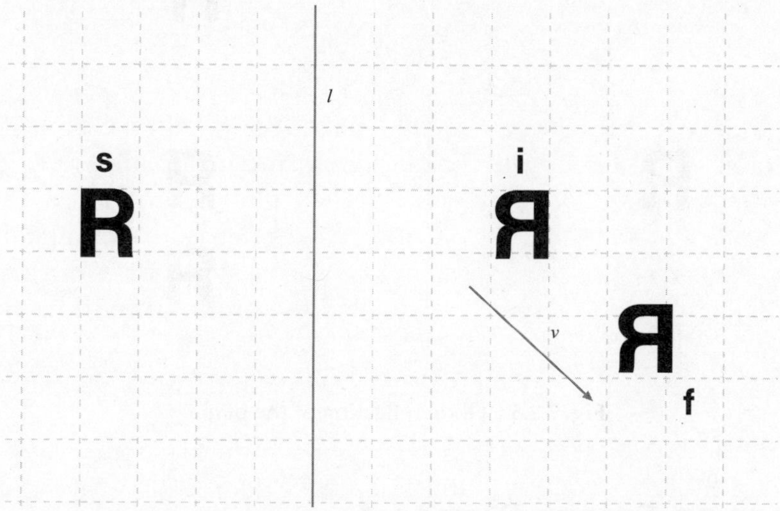

Fig. 7.2.4 Reflection followed by a translation of the plane

Concentrate on the start and finish position. You will note that no single rotation, reflection or translation of the plane can give us the same result. So we need to find a new symmetry of the plane which will do the needful. We are now in a position to define a glide reflection. A glide reflection is a symmetry of the plane where a reflection about a line *l* is followed by a translation about a vector *v* which is parallel to *l*. Figure 7.2.5 created using GeoGebra shows the glide reflection which gives us the same result as in Fig. 7.2.4. It shows the equivalent glide reflection which involves reflection about a line *g* followed by translation about a vector *u* which is parallel to *g*.

Thus symmetries of the plane are of four types—reflection, rotation, translation and glide reflection. The set of all symmetries of a plane is a group with infinitely many elements and is called the group of rigid motions of the plane. Formally, for any subset *X* of the plane, we can define a symmetry to be a distance preserving bijective function on the set *X* or alternatively as the collection of those symmetries of the plane that map *X* to itself. The '$\circ$' which denoted the application of one symmetry after the other is nothing but the composition of functions or maps.

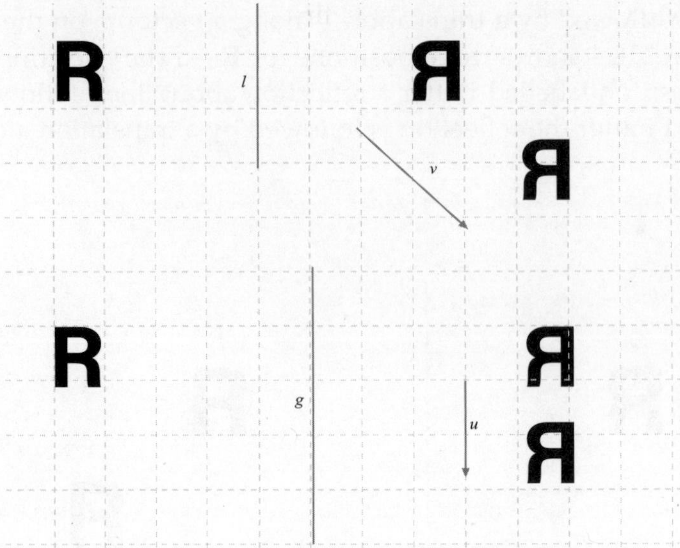

Fig. 7.2.5 Glide reflection of the plane

7.3 Strip or Frieze Patterns

Let us explore some infinite planar figures. The first of these is a strip or a frieze pattern. A frieze is essentially a horizontal band of decoration that is used in walls or ceilings. For us, a strip pattern or a frieze pattern will mean a finite planar figure, called motif, repeated at equal intervals indefinitely to the right and left to create an infinite horizontal strip of frieze. It is easy to find examples of strip patterns in decorations on monuments, sari borders, curtains, and in art and architecture. In this section, we will explore the symmetries of strip patterns. As seen at the end of the last section, a symmetry of the given strip pattern X will be a symmetry of the plane which maps X onto itself. Consequently the only symmetries that a strip pattern may have are reflections, rotations, translations and glide reflections.

We shall again use the intuitive definition of a symmetry that has served us so well. Given a strip pattern we need to identify those actions that we can perform on the strip pattern such that after the action it occupies the same position and looks exactly like it did before. Two things will certainly have a bearing on the symmetries of a strip pattern, one is the basic motif which is repeated and the other is the fact that the motif is repeated at equal intervals horizontally. Some of the constraints that are imposed by the latter are that if a strip pattern has a non-zero degree rotation then it has to be a 180 degree rotation. Similarly a strip pattern cannot have a reflection about a line which is oblique (neither horizontal nor vertical). On the other hand the horizontal repetition of the basic motif at equal intervals ensures that a strip pattern always has translation along a horizontal vector whose magnitude is the distance between two consecutive motifs. If the motif has reflection symmetries that are about a horizontal line or a vertical line, then one can expect the strip pattern to have similar symmetries. Also unless the basic motif has a 180 degree rotation symmetry it is unlikely that the strip pattern will have one.

Let us use GeoGebra to create and analyse the symmetries of a strip pattern. Figure 7.3.1 shows a strip pattern created using GeoGebra. The basic motif here is a square (created using the regular polygon option), and then repeated at equal intervals along the X-axis.

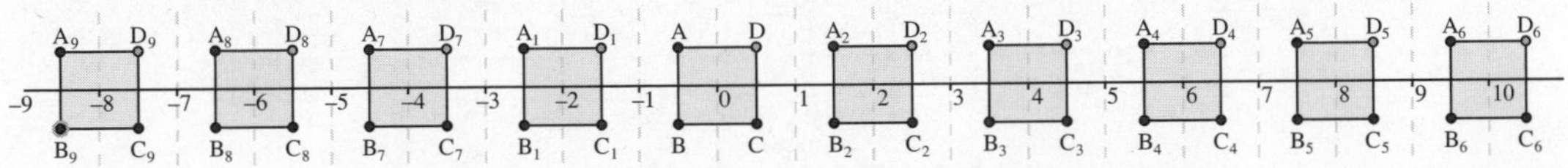

Fig. 7.3.1 Strip pattern with a square as a basic motif

The same strip pattern as in Fig. 7.3.1 is shown in Fig. 7.3.2 without any labels or axis.

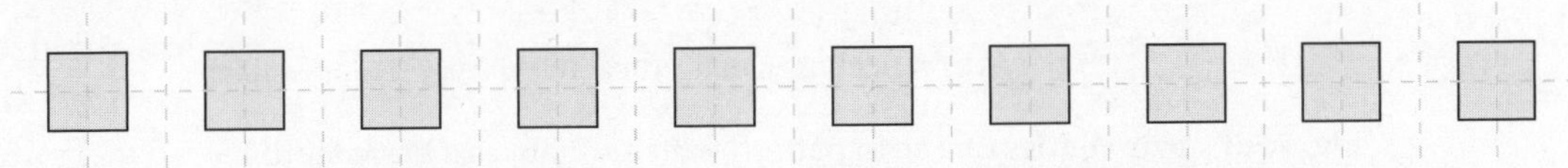

Fig. 7.3.2 Strip pattern with a square as a basic motif and no labels

The square, which was the basic motif in the above strip pattern, has eight symmetries. Four of these are rotations and the other four are reflections. The rotational symmetries are anticlockwise rotations about the centre (point of intersection of diagonals) of the square of degrees 0, 90, 180 and 270. Due to constraints imposed by being in a strip pattern, only the possibility of the 180 degree rotational symmetry inducing a symmetry of the strip pattern remains. Of course, every strip pattern will always have the 0 degree rotation. Similarly the two reflections about the diagonals will not give us any symmetries of this strip pattern but one can hope that the reflections about the horizontal line joining mid-point of opposite sides and similarly the vertical line might induce reflections of the strip pattern.

Recall that by the way a strip pattern is constructed it will always have translation symmetry and hence infinitely many translations in its group of symmetries. Figures 7.3.3 and 7.3.4 show the different types of symmetries that the strip pattern in Fig. 7.3.1 has. Note that the labelling of each square and the position it occupies is done to track its symmetries. `GeoGebra` has been used to perform the symmetries on the original strip pattern and then screen shots have been taken and assembled together to create the figures.

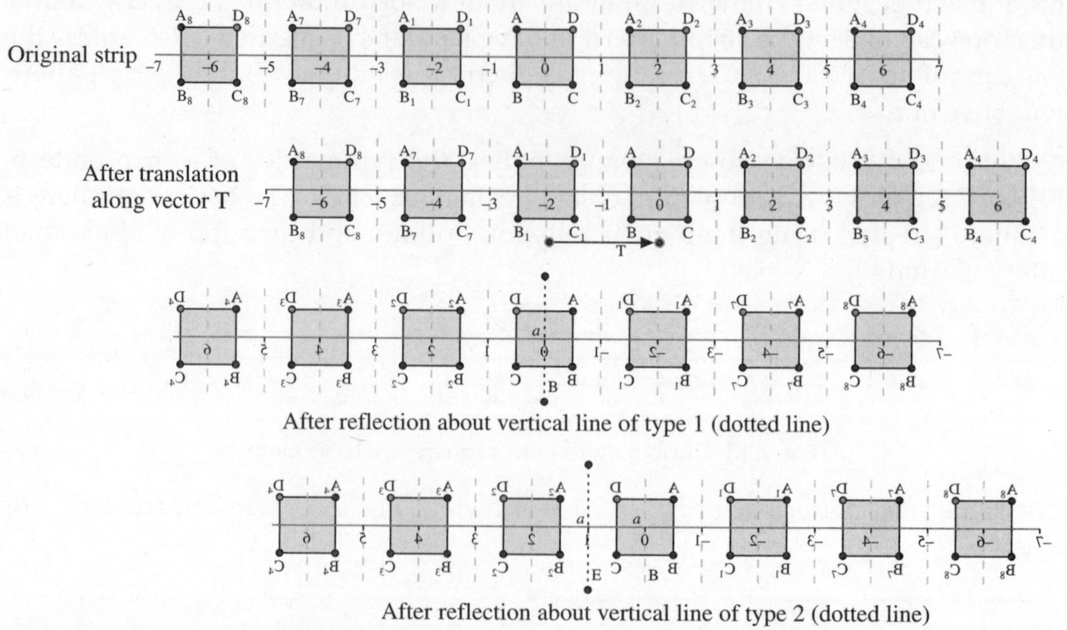

Fig. 7.3.3 Symmetries of a strip pattern with a square as a basic motif I

In Figs 7.3.3 we have shown translation of the strip along a vector T which gives us a symmetry as the vector T is parallel to the strip and has a magnitude equal to the distance between two consecutive motifs. If we denote this symmetry by t, then we also have the symmetries $t^2, t^3, \ldots$ where t^2 denotes the translation t applied twice or translation by the vector $2T$ and so on. We will also have $t^{-1}, t^{-2}, \ldots$ where t^{-1} denotes translation by the vector $-T$. Thus, this strip pattern and indeed any strip pattern will always have infinitely many symmetries coming from the collection $\{t^m \mid m \in \mathbb{Z}\}$. The symmetry t^0 is the same as the 0 degree rotation.

The other symmetry shown is reflection about two types of vertical lines. The vertical line of reflection of type 1 goes through the centre of the square, in fact, it is a line of reflection symmetry for that particular square. The other vertical line of reflection of type 2 goes through the mid point between two consecutive squares on the horizontal axis. Further, no other vertical lines on the strip except for these two types are lines of reflection symmetry.

Two lines of reflection are of the same type if a symmetry of the strip pattern takes one to the other. It is clear that no symmetry of this strip pattern can take a line of type 1 to that of type 2 or vice versa. On the other hand, there are infinitely many lines of reflection of each fixed type and the translation symmetries will take one to the other. In this strip pattern we also have infinitely many reflection symmetries about vertical lines of type 1 and of type 2.

Figure 7.3.4 shows the other type of symmetries that are possible for such a strip pattern. There is only one symmetry which is a reflection about the horizontal dotted line shown. However, when we consider 180 degree rotation of the strip pattern, there are two types of rotocenters that gives us rotational symmetries. Type 1 rotocenters are situated at the centre of a square motif. The type 2 rotocenters are at the mid-points between two consecutive squares on the horizontal axis. Further, no other points on the strip pattern gives us rotational symmetries.

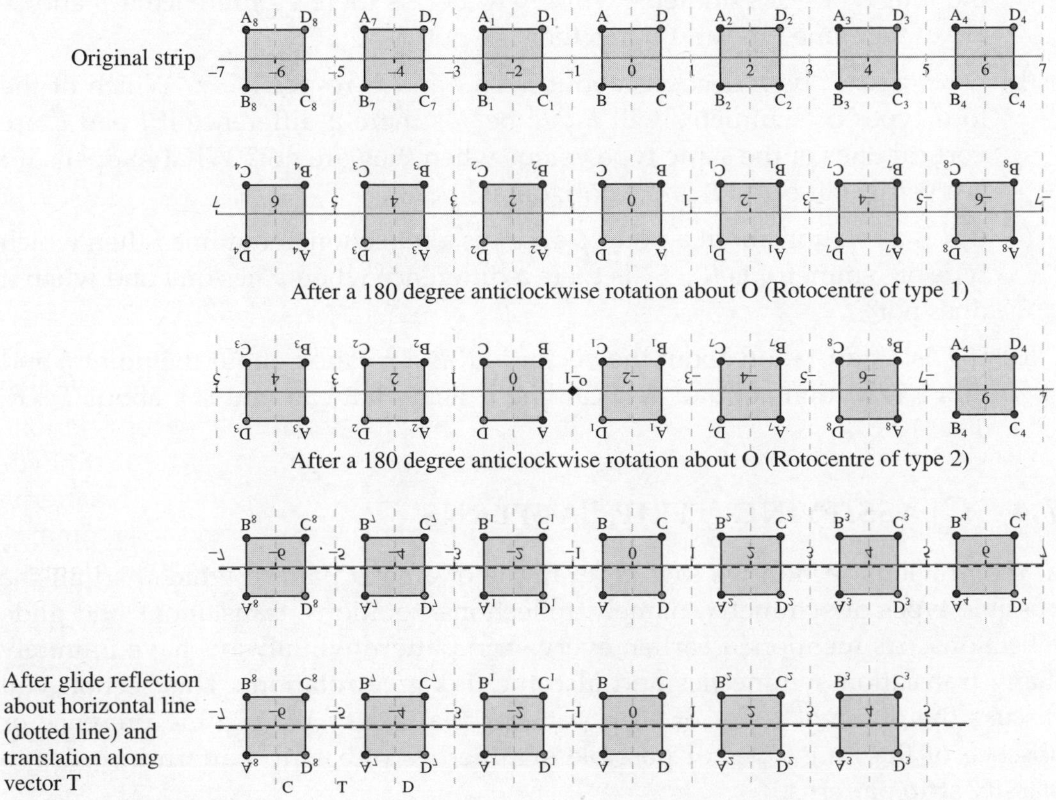

Fig. 7.3.4 Symmetries of a strip pattern with a square as a basic motif II

Two rotocenters are of the same type if a symmetry of the strip pattern takes one to the other. It is clear that no symmetry of this strip pattern can take a rotocenter of type 1 to a rotocenter of type 2. Further translation symmetry ensures that two rotocenters of the same type can be reached. We will have infinitely many rotational symmetries of this strip pattern as there are infinitely many rotocenters of each type.

The last type of symmetry shown in Fig. 7.3.4 is that of a glide reflection which encapsulates a reflection about the horizontal axis followed by the translation t. There will be infinitely many glide reflections as the translation can happen along the vectors nT as n varies over the integers.

Investigation 7.4 Use `GeoGebra` to explore the questions below about symmetries of the strip pattern in Fig. 7.3.1. Again we assume that the symmetries discussed below are non-trivial.

(a) Let R and R' be rotations about rotocenters p and p' respectively. Which of the four types of symmetry will $R \circ R'$ be? Is there a difference if p and p' are of the same type and when they are not?

(b) Let F and F' be reflections about lines l and l' respectively. Which of the four types of symmetry will $F \circ F'$ be? Is there a difference if l and l' are vertical lines of the same type versus when they are not? What happens if l is a vertical line and l' is the horizontal line?

(c) If R is a rotation about a point p and F is a reflection about line l then which type of symmetry is $R \circ F$? Is there a difference when p lies on l and when it does not?

(d) If T is a translation about the vector v, R is a rotation about the point p and F is a reflection about a vertical line l then what can you say about $T \circ R$, $T \circ F$?

7.4 Classifying Strip Patterns

In the previous section we saw an example of a strip pattern which had all the possible types of symmetry, namely, reflections, rotations, translations and glide reflections. As mentioned earlier, every strip pattern will always have infinitely many translation symmetries and also the 0 degree rotation. This section will discuss the different types of strip patterns that exist. We use the presence or absence of the four types of possible symmetries of a strip pattern in order to classify strip patterns.

There are seven types of strip patterns possible and examples of each type are given below. Later in this section we will discuss the method or algorithm which will help us classify strip patterns. Let us denote these seven patterns by SP Pattern I–VII. In all the types mentioned below, translation symmetry and 0 degree rotational symmetry will always be there even if not mentioned explicitly. All the strips have been made using `GeoGebra`.

SP Pattern I consists of strip patterns that only have translation symmetry (Fig. 7.4.1).

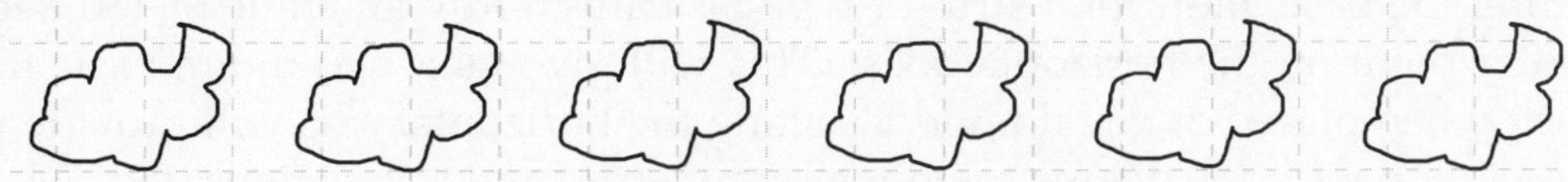

Fig. 7.4.1 Strip pattern of type SP Pattern I

The second type, SP Pattern II, has strip patterns that have only glide reflections and translation symmetry. You might wonder how to create such strips. Choose the same basic motif as in SP Pattern I, reflect it about the horizontal axis and translate it by a vector g along the horizontal axis. This new two figure motif will serve as the basic motif for the strip patterns of type SP Pattern II. This is illustrated in Fig. 7.4.2. The glide reflections here involve reflecting the strip about the horizontal dotted line and then translating along a vector $(2n + 1)g$ where n is an integer. The translation symmetries here are given by translation along a vector $2ng$ where n is an integer. Also note that this strip does not have a reflection about a horizontal line even though it has translations and glide reflections.

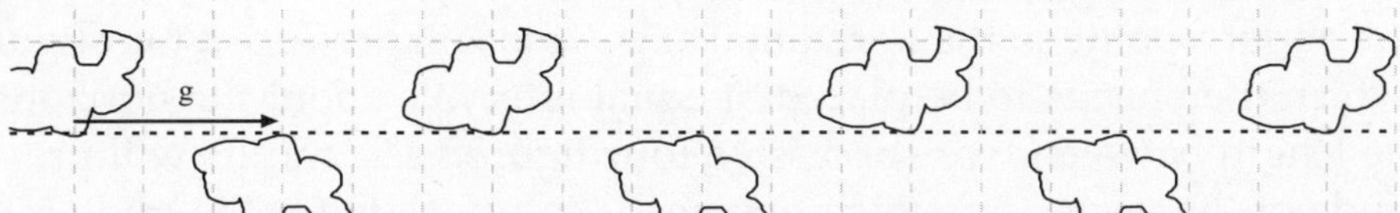

Fig. 7.4.2 Strip pattern of type SP Pattern II

SP Pattern III consists of strip patterns that only possess reflections about vertical lines and translations. Again we can construct a basic motif for strip patterns of this sort by reflecting the basic motif of a strip of type SP Pattern I about a vertical line. This will give us a two figure basic motif which when placed at equal intervals along the horizontal axis will provide us a strip pattern with two types of lines of vertical symmetry and translations.

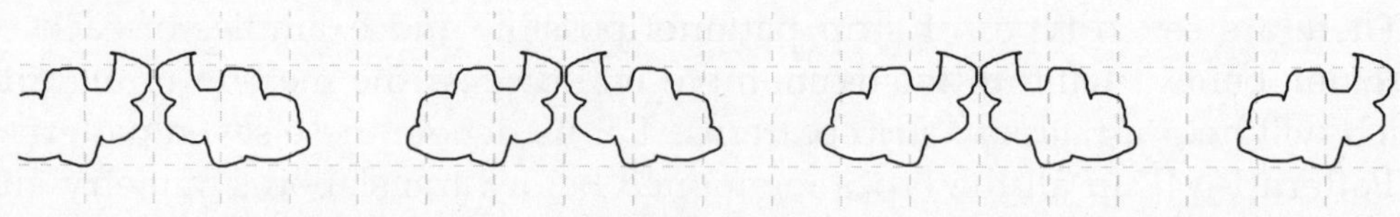

Fig. 7.4.3 Strip pattern of type SP Pattern III

SP Pattern IV consists of strip patterns that only possess 180 degree rotations and translations. Again we can construct a basic motif for strip patterns of this sort by rotating the basic motif of a strip of type SP Pattern I by an angle of 180 degree about a point on the horizontal axis. This will give us a two figure basic motif which when placed at equal intervals along the horizontal axis will provide us a strip pattern with 180 degree rotational symmetries and translations (Fig. 7.4.4).

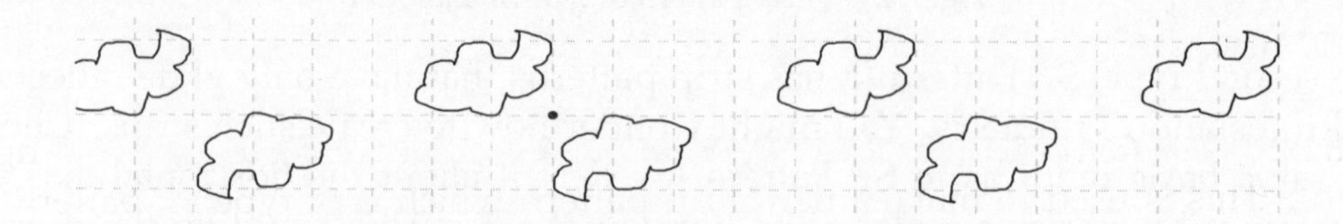

Fig. 7.4.4 Strip pattern of type SP Pattern IV

The next pattern type, SP Pattern V has strip patterns that only possess glide reflections, reflections about vertical lines, 180 degree rotations and translations. This time we create a four figure basic motif. Start with a basic motif like that in Fig. 7.4.1. Then reflect it about a vertical line. Now we have a two figure motif. Now rotate this two figure motif by an angle of 180 degree in the counter-clockwise direction about a point on the horizontal axis. This will produce a four figure basic motif which then needs to be placed at equal intervals about the horizontal axis. Half the length between two successive four figure basic motifs will provide the magnitude of the vector by which one needs to translate for the glide reflection. This construction will give a strip pattern of type SP Pattern V (Fig. 7.4.5).

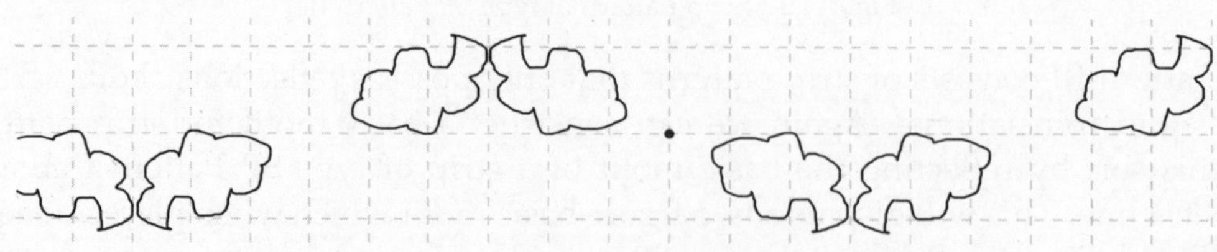

Fig. 7.4.5 Strip pattern of type SP Pattern V

SP Pattern VI has strip patterns that only possess horizontal reflection, glide reflections and translations as symmetries. The basic motif here is a two figure motif created by reflecting the basic motif of a strip of SP Pattern I about a horizontal line. Then this two figure motif is used to create a strip pattern. Such a strip will be of SP Pattern VI as seen in Fig. 7.4.6.

Fig. 7.4.6 Strip pattern of type SP Pattern VI

The final strip pattern is SP Pattern VII. This type of strip pattern has all four types of symmetry possible. We have seen an example of this already with the strip pattern with a square as a basic motif. We can also use the basic motif of a strip of SP Pattern I to create a four figure motif which will serve as a basic motif for the strip that we require. First reflect the basic motif about a vertical line to get two figures. Then reflect the two figures about a horizontal line to get a four figure motif. This is used to make a new strip pattern which is of type SP pattern VII. Figure 7.4.7 illustrates this.

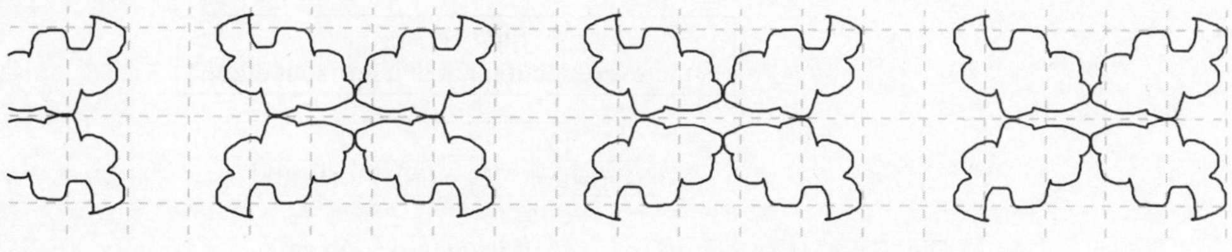

Fig. 7.4.7 Strip pattern of type SP Pattern VII

Below is a procedure or algorithm that will help classify a given strip pattern. It is based on a flow chart given in [15].

Step 1: Check if there is a reflection about a vertical line. If the answer is yes, then go to Step 2Y and if the answer is no, then go to Step 2N.

Step 2Y: Check if there is a reflection about a horizontal line. If the answer is yes, then it is a strip of pattern SP Pattern VII. If the answer is no, then go to Step 2Y3N.

Step 2Y3N: Check if there is a 180 degree rotation. If the answer is yes, then it is of SP Pattern V. If the answer is no, then it is of SP Pattern III.

Step 2N: Check if there is a glide reflection or a horizontal reflection. If the answer is yes, then go to Step 2N3Y, and if the answer is no, then go to Step 2N3N.

Step 2N3Y: Check if there is a reflection about a horizontal line. If the answer is yes, then it is a strip of pattern SP Pattern VI. If the answer is no, then it is a strip of pattern SP Pattern II.

Step 2N3N: Check if there is a 180 degree rotation. If the answer is yes, then it is of SP Pattern IV. If the answer is no, then it is of SP Pattern I.

The above steps are illustrated in the chart shown in Fig. 7.4.8.

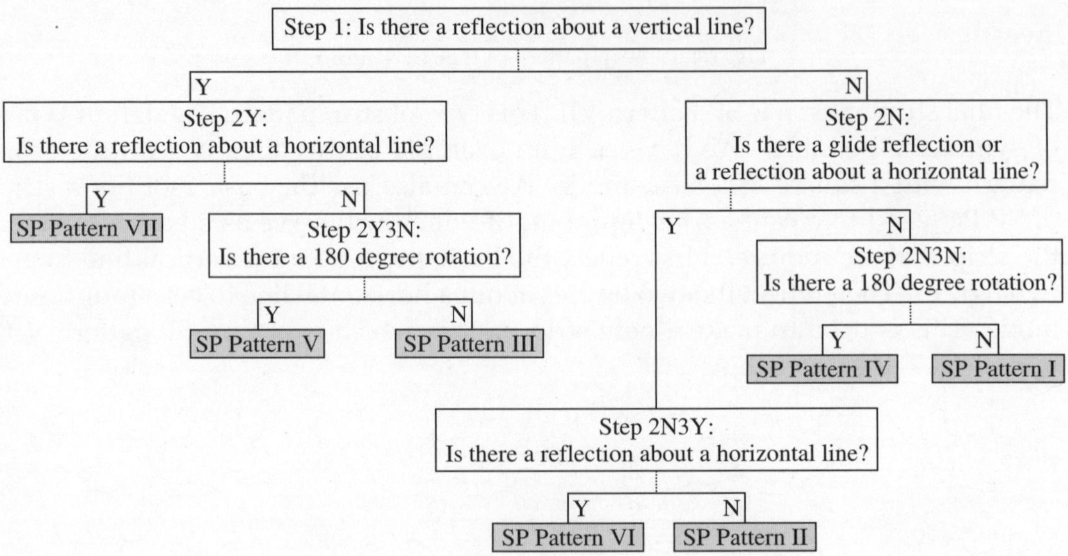

Fig. 7.4.8 Procedure for classifying strip patterns

Now let us try the above procedure in the context of the test strip pattern shown in Fig. 7.4.9.

Fig. 7.4.9 Test strip pattern

It is clear that there is no reflection about a vertical line. So we move to **Step 2N** and ask if there is a glide reflection or a horizontal reflection. The answer is again no. This can be checked also using `GeoGebra` by importing the image and reflecting about the horizontal line between the upright and upside down trees. Therefore we move to **Step 2N3N** and ask if there is a 180 degree rotation. Using `GeoGebra` we see that indeed there is 180 degree rotation about a point on the horizontal bisector placed midway between an upright and upside down tree. So by our method the test strip is of type SP Pattern IV.

Investigation 7.5 Use `GeoGebra` to explore the questions below.

(a) Create a basic motif which has only the 0 degree rotational symmetry and use it to create strips of SP Patterns I–VII.

(b) Use the step wise method for finding which pattern the strips given in Figs 7.4.10 and 7.4.11 satisfy.

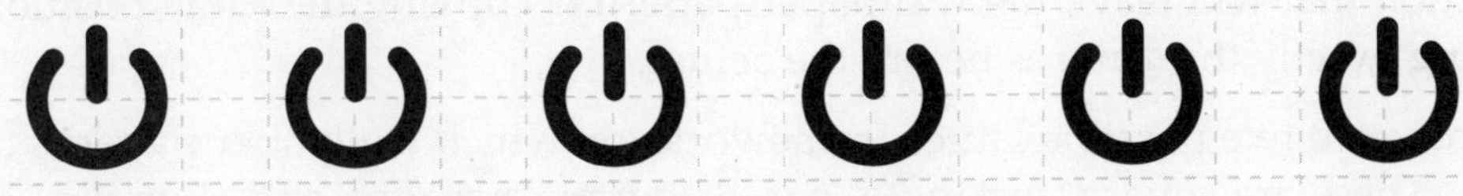

Fig. 7.4.10 Test strip pattern (i)

Fig. 7.4.11 Test strip pattern (ii)

7.5 WALLPAPER PATTERNS

Another infinite planar figure that gives rise to interesting combinations of symmetries is a wallpaper pattern. An easy way to create a type of wallpaper pattern is to take a strip pattern and then to stack it at equal intervals in the vertical direction. In a strip pattern, both the basic motif and the repetition in the horizontal direction influence the symmetries of the pattern. In a wallpaper pattern created by stacking copies of a strip pattern, equally spaced in the vertical direction, it is to be expected that the symmetries of the strip pattern as well as the infinite repetition in the left-right and up-down direction will determine the symmetries of a wallpaper pattern.

In general, a wallpaper pattern is a periodic pattern that covers the plane. In order to construct or alternatively deconstruct a wallpaper pattern, we need to first understand the possible framework or lattice that is like the skeleton for the wallpaper pattern. Think of these as dots placed in a particular geometric configuration which is periodically repeated to cover the plane. We can then replace the dots by a basic motif to create a wallpaper pattern. Wallpaper patterns can be found in the window decorations or 'jalis' of old monuments. Examples can also be found in carpets, wallpapers, floor and wall tiles among others.

In this section we will explore the underlying framework or lattices along whose points we place basic motifs to get wallpaper patterns. We also explore the symmetries that will always exist for a wallpaper pattern and consider several examples. Most importantly we will use `GeoGebra` to construct wallpaper patterns and to experiment with the different symmetries that the pattern may or may not have. Since the symmetry of a given wallpaper pattern is a symmetry of the plane that takes the wallpaper pattern back to itself, it can only be either a translation, a reflection, a rotation or a glide reflection. Also we can think of a symmetry as an action that will leave the wallpaper pattern occupying the same position and looking exactly the same as before the action.

There are five possible lattice frameworks on which wallpaper patterns can be built. These are square, rectangle, parallelogram, hexagon and rhombic. Figure 7.5.1 illustrates each of these 5 types of lattices and have been created using `GeoGebra`.

Figure 7.5.2 shows wallpaper patterns corresponding to the 5 types of lattices arranged in the same order as in Fig. 7.5.1. These wallpaper patterns have been created using the wallpaper symmetry generating application freely available at [80].

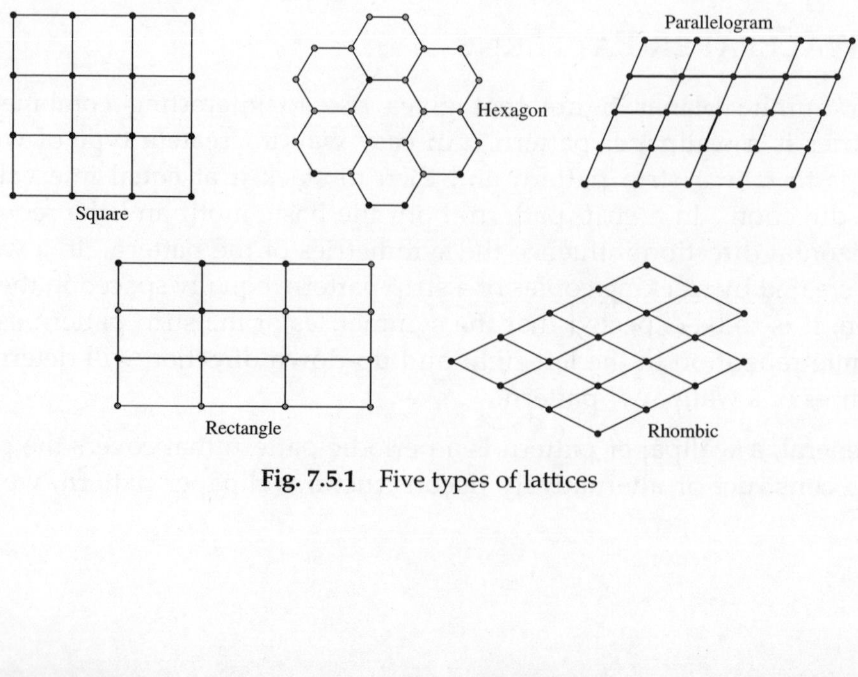

Fig. 7.5.1 Five types of lattices

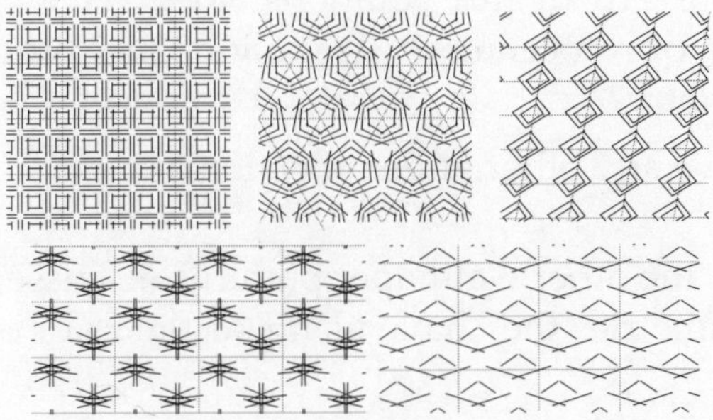

Fig. 7.5.2 Wallpaper patterns based on the 5 types of lattices

By the nature of their construction, wallpaper patterns will have translations in two independent directions. They may or may not have lines of reflection symmetry. Here we do not have to restrict ourselves to just vertical or horizontal lines. There may also be oblique lines of reflection. Due to the underlying lattices, the smallest non-zero degree rotations may be only from 60, 90, 120 or 180 degree. A wallpaper pattern may or may not have glide reflections.

Figure 7.5.3 illustrates some of the symmetries for the first wallpaper pattern in Fig. 7.5.2. These have been created using `GeoGebra`.

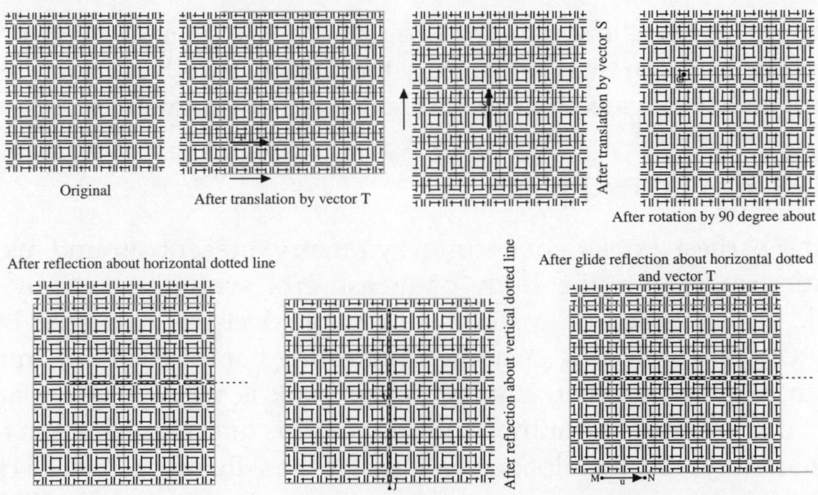

Fig. 7.5.3 Symmetries of a wallpaper pattern based on a square lattice

Investigation 7.6 For the wallpaper pattern (WPSQ) discussed in Fig. 7.5.3, answer the following using `GeoGebra` or otherwise.

(a) Does WPSQ have any other types of lines of reflection other than those shown in the above figure?

(b) What other non-zero degree rotations does WPSQ have? Does it have any other types of rotocenters? Are any of the rotocenters on the lines of reflection symmetry?

(c) Does WPSQ have any other glide reflection along a non-horizontal axis?

Investigation 7.7 For the other wallpaper patterns discussed in Fig. 7.5.3, find the possible types of symmetries they have using `GeoGebra` or otherwise.

Based on the combination of symmetries that the wallpaper pattern possesses, they can be classified into 17 different types. The method for the classification first asks for the smallest non-zero degree rotation that the wallpaper possesses. As mentioned earlier, due to the underlying lattices this can only be $60, 90, 120$ or 180 degree. The next question asks if there are any reflections for the wallpaper pattern. If the answer is yes then depending on the value of the smallest rotation degree, questions can range from whether there are glide reflections or not, if the glide axis is the same as a reflection axis, whether there are two or more directions for the reflection axis and whether rotocenters lie on lines of reflection and so on. This is a lot harder than classifying strip patterns. For further details, the interested reader may refer to the chapter dealing with crystallographic groups in [15].

Investigation 7.8 Use the symmetry app referred to earlier, to create a wallpaper pattern. Take a screen shot and insert the image into `GeoGebra`. Explore the wallpaper you have created for the different types of symmetries.

FURTHER EXPLORATIONS

One of the earliest books connecting symmetry present around us with the mathematics underlying the study or measure of symmetry was by Hermann Weyl [83]. It was based on a set of four public lectures delivered by Weyl in Princeton University in 1951. Another interesting book which any informed lay person can read on symmetry is [10]. The book is written as a mathematical journey of discovery by the author and contains his quest with his son for finding the 17 different types of wallpaper patterns at the Alhambra Palace [1]. You can also read a review of this book and other essays on symmetry in [75], [76] and [77]. For an intuitive treatment of planar symmetries, a case study of Humayun's Tomb via symmetry and an exploration of Escher-like art through symmetry see [3]. Another direction in which symmetry could be explored is by considering symmetries of three dimensional finite objects, in particular the platonic solids. See [13] for an exploration of the symmetries of platonic solids using `GeoGebra`.

8

Applications of Matrices

A matrix is defined as a set of numbers arranged in rows and columns. It is a tool for both organising and processing data. J J Sylvester [57] first introduced the term 'Matrix Theory' in 1848 and the algebra of matrices was further developed by Arthur Cayley [42] in 1855. Matrix theory or linear algebra has wide applications in various fields such as economics, physics and engineering as well as in different branches of mathematics. In fact, matrices and linear algebra provide important tools for mathematical modelling. In this chapter we shall explore applications of matrix theory to ranking players or teams in a tournament, to the inheritance of traits in genetics and to an age structured population growth model. `MS Excel` and `Mathematica` will be the primary vehicles of exploration in the investigations.

8.1 Ranking Players

Let us consider the following situation. Fours players A, B, C and D participate in a tennis tournament where each player plays against the other. The results are as follows: A defeats B and C, B defeats D, C defeats B and D, D defeats A. You have been assigned the task of ranking the players. How would you go about it?

We need some guiding principles. For example, the most obvious criterion is to look at who won the most matches. Then A and C are the best players, while B and D are worst. But how do we distinguish between A and C, or between B and D? One approach could be to rank A above C because A beat C, and similarly rank B above D. This approach will generally break ties, unless we run into the following situation.

Investigation 8.1 The results of a 3-player tournament were as follows: A beat B, B beat C, C beat A. Is there any way of ranking the players?

Suppose we come across a situation such as the one in the above investigation. If other players were also participating, then you could look into whom A, B and C beat. Did A beat better players than the ones that B beat? If yes, you have reason to rank A above B. It is not obvious how to carry out such an evaluation. In this section, you will see how graph theory and matrices can help.

Directed Graphs

We shall represent the results of the tennis tournament in the form of a directed graph. By a *directed graph* we mean a finite set V of elements $v_1, v_2, \ldots, v_n$ together with a finite collection E of ordered pairs (v_i, v_j) of distinct elements of this set where no ordered pair is repeated. The elements of V are called *vertices* and the members of E are called *directed edges* of the directed graph. The notation $v_i \to v_j$ (read as v_i is connected to v_j) is used to indicate the fact that the ordered pair (v_i, v_j) belongs to the directed graph. Graphically we may represent the vertices v_i as points in the plane and $v_i \to v_j$ by drawing a line or arc from v_i to v_j with an arrow pointing from v_i to v_j. Further, if the ordered pair (v_i, v_j) belongs to the directed graph then the reversed ordered pair (v_j, v_i) does not belong to the directed graph.

A *dominance directed graph* is a directed graph in which every pair of vertices is joined by a directed edge. If $v_i \to v_j$ is a directed edge of a dominance directed graph, we say that v_i *dominates* v_j.

Let us represent the results of the tennis tournament as a directed graph. The four players A, B, C and D may be represented as vertices of the graph. If a player has defeated another player we will join the two vertices (representing the two players) with a directed edge (a line with an arrow pointing towards the player who has been defeated). Since all players play each other, the directed graph shown in Fig. 8.1.1 is a dominance directed graph.

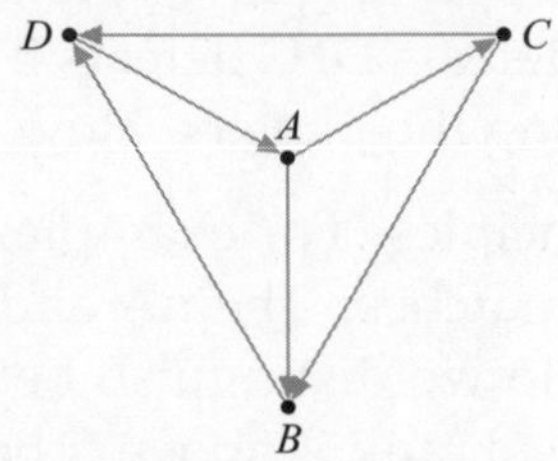

Fig. 8.1.1 A dominance directed graph representing the wins in the tennis tournament

Matrix Representation of Directed Graphs

With a directed graph of n vertices, we associate an $n \times n$ matrix $M = [m_{ij}]$ called the *vertex matrix* or *dominance matrix* whose elements are defined by

$$m_{ij} = \begin{cases} 1 & \text{if } v_i \to v_j, \\ 0 & \text{else.} \end{cases} \qquad \text{for } i, j = 1, 2, \ldots, n.$$

Dominance matrices, by their very definition, have the following properties:

(i) All the entries are either 0 or 1.

(ii) If $m_{ij} = 1$ then $m_{ji} = 0$.

(iii) All diagonal entries are 0.

Using the above definition, the vertex matrix of the directed graph in Fig. 8.1.1 is

$$M = \begin{bmatrix} 0 & 1 & 1 & 0 \\ 0 & 0 & 0 & 1 \\ 0 & 1 & 0 & 1 \\ 1 & 0 & 0 & 0 \end{bmatrix}$$

The first, second, third and fourth rows (and columns) of the matrix M correspond to the players A, B, C and D respectively. Thus $m_{ij} = 1$ indicates a win of player i over player j. Also, when $i \neq j$, then $m_{ij} = 0$ indicates that player i has lost to player j. In our tennis tournament, since A has defeated B, $m_{12} = 1$. However A has lost to D and therefore $m_{14} = 0$.

Investigation 8.2 Represent the directed dominance graph of Fig. 8.1.2 by a dominance matrix.

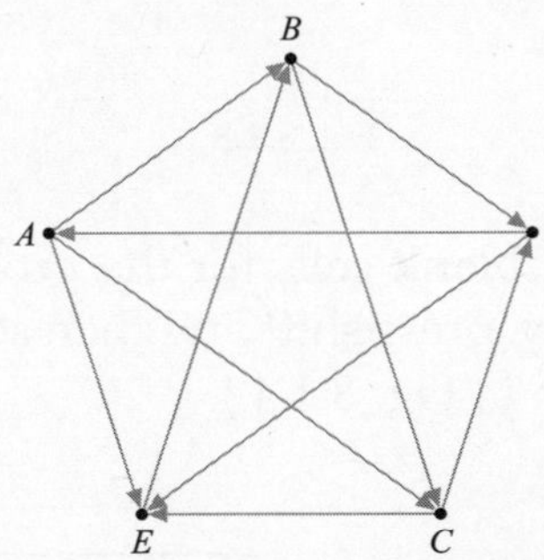

Fig. 8.1.2 A directed dominance graph with 5 vertices

Investigation 8.3 Draw the directed graph corresponding to the dominance matrix given below. Is it a dominance directed graph?

$$\begin{bmatrix} 0 & 1 & 0 & 1 & 1 \\ 0 & 0 & 1 & 0 & 1 \\ 1 & 0 & 0 & 1 & 1 \\ 0 & 1 & 0 & 0 & 1 \\ 0 & 0 & 0 & 0 & 0 \end{bmatrix}$$

Coming back to our tennis tournament, we have already seen that the 4×4 dominance matrix M represents the wins and losses of each player. A *row sum* is the sum of all the entries in a row. Each row sum equals the number of wins of the corresponding player. If we assign the row sums as the scores for the corresponding players, then players A and C get a score of 2 each while B and D get 1. However, this does not help us rank the players completely.

Let us see what information may be obtained from the matrix M^2, which is the product of the matrix M with itself. First, let us learn how to use `Excel` to multiply matrices.

Matrix Computation in Excel

For computing M^2 we may use a spreadsheet. `Excel` provides the `MMULT` command which allows us to multiply matrices. The steps of matrix multiplication in Excel are as follows:

Step 1: The entries of matrix M are entered as a 4×4 array as shown in Fig. 8.1.3.

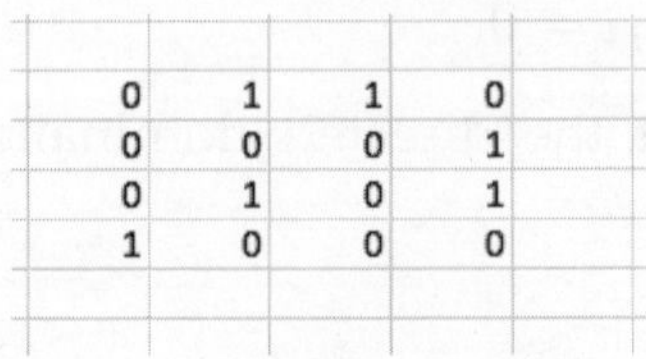

Fig. 8.1.3

Step 2: Select a 4×4 array of blank cells for the product matrix M^2. (Select the top left corner of the array, press shift and then select the bottom right corner cell of the array as shown in Fig. 8.1.4.)

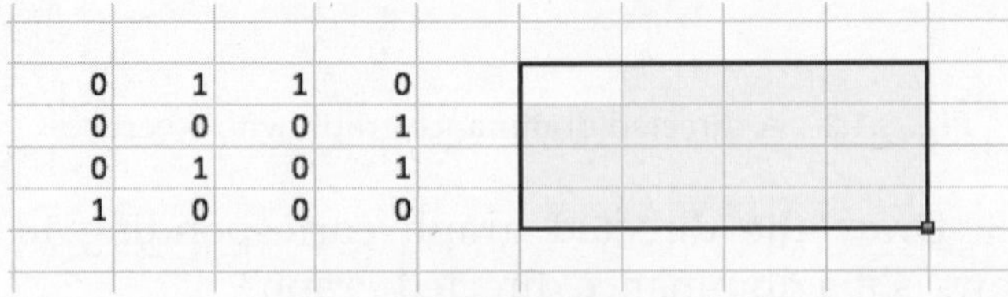

Fig. 8.1.4

Step 3: In the top left corner cell of the selected array, enter `=MMULT(`. This is shown in Fig. 8.1.5. Note that `Excel` indicates that array 1 and array 2 have to be selected (array 1 and array 2 are the two matrices whose product is required).

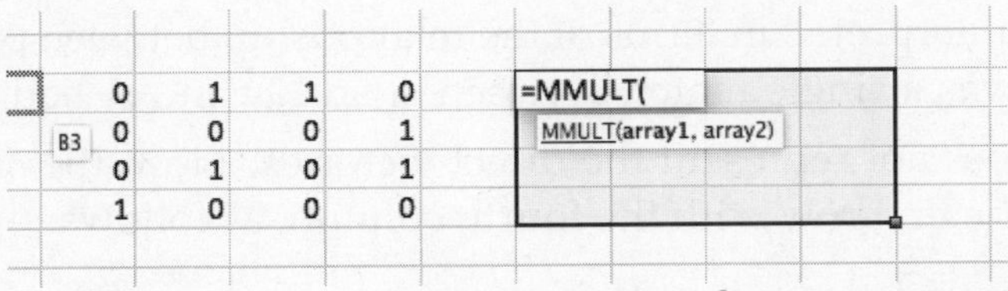

Fig. 8.1.5

Step 4: Within the `MMULT` command select the array containing the entries of matrix M. Place a comma and select the array again. (Here array 1 and array 2 are the dominance matrix M as shown in Fig. 8.1.6.)

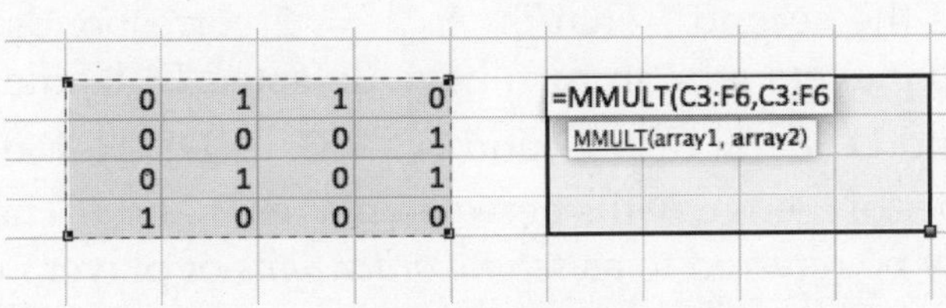

Fig. 8.1.6

Step 5: Press the keys `SHIFT + CRTL + ENTER` to obtain M^2. See Fig. 8.1.7.

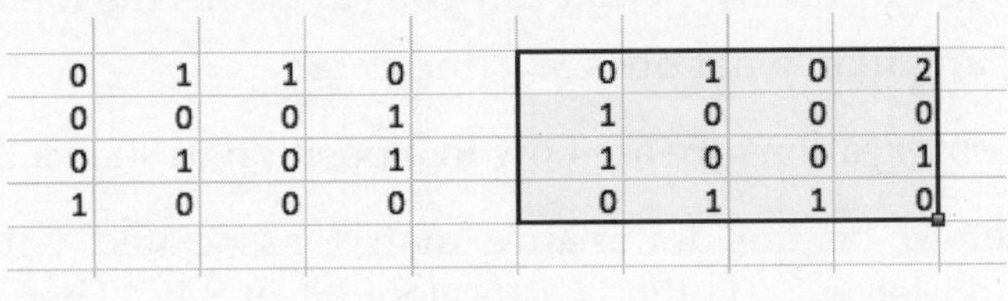

Fig. 8.1.7

This gives us

$$M^2 = \begin{bmatrix} 0 & 1 & 1 & 0 \\ 0 & 0 & 0 & 1 \\ 0 & 1 & 0 & 1 \\ 1 & 0 & 0 & 0 \end{bmatrix} \begin{bmatrix} 0 & 1 & 1 & 0 \\ 0 & 0 & 0 & 1 \\ 0 & 1 & 0 & 1 \\ 1 & 0 & 0 & 0 \end{bmatrix} = \begin{bmatrix} 0 & 1 & 0 & 2 \\ 1 & 0 & 0 & 0 \\ 1 & 0 & 0 & 1 \\ 0 & 1 & 1 & 0 \end{bmatrix}$$

Using the Dominance Matrix to Rank Players

Notice that M^2, like M, has 0's and 1's as its entries, but in addition it also has 2 as an entry. We may refer to the entries of M^2 as $m_{ij}^{(2)}$. But what does $m_{ij}^{(2)}$ represent in terms of the wins and losses of the players i and j? Let us observe that $m_{14}^{(2)} = 2$.

How can this be interpreted in terms of the relationship between players A and D? Note that there was a single match between A and D and A had lost to D. Thus, the matrix M^2 does not represent the direct victories. Now, the $m_{14}^{(2)}$ entry comes by multiplying the first row with the fourth column, as follows:

$$\begin{bmatrix} 0 & 1 & 1 & 0 \end{bmatrix} \begin{bmatrix} 0 \\ 1 \\ 1 \\ 0 \end{bmatrix} = 0 \cdot 0 + 1 \cdot 1 + 1 \cdot 1 + 0 \cdot 0 = 1 + 1 = 2$$

The $A \to B$ and $B \to D$ edges create the first $1 \cdot 1$ in the product, while $A \to C$ and $C \to D$ create the second. That is, $m_{14}^{(2)} = 2$ corresponds to the fact that A has defeated two players who in turn have defeated D, leading to two 2-edge paths between A and D ($A \to B \to D$ and $A \to C \to D$). Hence the entries $m_{ij}^{(2)}$ represent the number of 2-edge paths between players i and j. These 2-edge paths between i and j may be referred to as *second order wins* of player i over player j.

Investigation 8.4 Identify the second order wins among the players A, B, C and D represented by the non-zero entries of the matrix M^2.

Investigation 8.5 Compute M^3. What can you say about the interpretation of the entries $m_{ij}^{(3)}$? In particular, what does $m_{23}^{(3)}$ indicate?

Based on our observations we are now in a position to state a theorem.

Theorem 8.1 Let M be the dominance matrix associated with a dominance directed graph and let $m_{ij}^{(r)}$ be the (i, j)th element of M^r. Then $m_{ij}^{(r)}$ is equal to the number of r-edge paths from vertex v_i to v_j.

The row sums of M^2 give us additional information about the players. The row sum of A will be higher if the players beaten by A beat more players themselves, that is, if A beats stronger players. In our example, the M^2 scores for A, B, C and D are now 3, 1, 2 and 2 respectively. We may use these scores along with the row sum scores of M to rank the players. In fact the combined scores of M and M^2 may be obtained using the matrix operation of addition as follows.

$$M + M^2 = \begin{bmatrix} 0 & 1 & 1 & 0 \\ 0 & 0 & 0 & 1 \\ 0 & 1 & 0 & 1 \\ 1 & 0 & 0 & 0 \end{bmatrix} + \begin{bmatrix} 0 & 1 & 0 & 2 \\ 1 & 0 & 0 & 0 \\ 1 & 0 & 0 & 1 \\ 0 & 1 & 1 & 0 \end{bmatrix} = \begin{bmatrix} 0 & 2 & 1 & 2 \\ 1 & 0 & 0 & 1 \\ 1 & 1 & 0 & 2 \\ 1 & 1 & 1 & 0 \end{bmatrix}$$

The combined row sum scores are 5, 2, 4 and 3 for players A, B, C and D respectively. Since A has the largest row sum, it has the largest number of 1-edge and 2-edge connections to other vertices. In this sense, A is the most dominant among all the four vertices. In fact, with this more refined look at the dominance relationships, we can say that A is the strongest player, followed by C, D, B, in that order.

Let M be the dominance matrix of a dominance directed graph. The *power* of a vertex of the graph is the total number of 1-edge and 2-edge paths from it to other vertices. Alternatively, the power of a vertex v_i, is the ith row sum of the matrix $P = M + M^2$.

We may also assign weightages to the first order and second order victories using the matrix equation $\alpha M + \beta M^2$ where α and β are scalar quantities between 0 and 1. By varying α and β we can adjust the relative importance of direct wins and second order wins.

Investigation 8.6 Use the matrix equation $\alpha M + \beta M^2$ with $\alpha = 0.7$ and $\beta = 0.3$ to rank the players A, B, C and D.

Let us also see how `Mathematica` can be used in such calculations. First, a matrix can be defined row-wise. For example, the dominance matrix M of the tennis tournament example can be defined as follows.

```
M={{0,1,0,1,1,0},{0,0,1,1,0,1},{1,0,0,0,1,1},
 {0,0,1,0,1,0},{0,1,0,0,0,1},{1,0,0,1,0,0}};
```

The following commands carry out all the required computations.

- View M formatted as a matrix: `M//MatrixForm`.
- Draw the corresponding directed graph: `AdjacencyGraph[M,VertexLabels->"Name"]`.
- Compute the nth power of M: `MatrixPower[M,n]`.
- Add matrices M and N: `M+N`.
- Compute the row sums: `Total[M,{2}]`.
- Compute the column sums: `Total[M,{1}]`.

These commands should make it easy for you to carry out the next investigation.

Investigation 8.7 In this investigation, we shall see that the ranking methods can be applied even when there are vertices that are not directly connected. (The case of a general directed graph rather than a dominance directed graph.) Consider a small airline, which operates flights between six cities. We will refer to the cities as C_i, where $i = 1, \ldots, 6$. The direct flights between the cities are as follows.

Flight originating from	Direct flight to
C_1	C_5, C_6
C_2	C_1, C_4, C_6
C_3	C_2, C_5
C_4	C_3, C_5, C_6
C_5	C_2
C_6	C_3, C_5

(a) Draw a directed graph to represent the flight connections between the six cities.

(b) Write the corresponding dominance matrix M and compute M^2 and M^3.

(c) Compute the row sums of the following matrices to rank the cities in terms of ability to travel from there to other places.

(i) M

(ii) $M + M^2$

(iii) $M + 0.1M^2$

(iv) $M + 0.1M^2 + 0.1M^3$

Investigation 8.8 Modify the data of the previous investigation as follows:

Flight originating from	Direct flight to
C_1	C_2, C_4, C_5
C_2	C_3, C_4, C_6
C_3	C_1, C_5, C_6
C_4	C_3, C_5
C_5	C_2, C_6
C_6	C_1, C_4

Are you able to rank the cities using combinations of M, M^2 and M^3? Or even with higher powers of M?

8.2 Genes and Inheritance

In this section we shall explore some applications of matrices to genetics. In particular we shall look at genetic models of inheritance. In any species, genetic traits are transmitted from one generation to the next through genes. Genes are in fact the fundamental units of heredity. In humans, characteristics such as height, colour of eye or hair, and blood type are transmitted from parents to their children. In plants, traits such as colour of flowers, seed type and texture of leaves are also transmitted from the parent plants to their offspring. Typically, each characteristic of an individual is determined by a pair of genes, one received from each parent. Each gene is either *dominant* or *recessive*. Dominant genes are represented using capital letters while lower case letters are used to represent the recessive genes. Let us consider a particular trait in a plant, such as flower colour. Let us assume that the dominant gene **A** leads to red colour and the recessive gene **a** leads to white colour in the flowers. These two genes lead to three possible *genotypes*, that is, **AA**, **Aa** and **aa**. The genotypes **AA** and **Aa** produce red flowers whereas **aa** leads to white flowers in the plant. Thus, in the hybrid genotype **Aa**, the **A** gene dominates over the **a** gene and the colour manifested is red. The actual trait manifested (in this case the colour of the flower) is referred to as the *phenotype*.

According to the rules of genetics, when two individuals mate, the offspring plant will inherit one gene from each parent plant with equal probability. Let us see what happens when two individuals, one with genotype **AA** and the other with genotype **Aa** mate. There are four possibilities:

(i) The offspring gets the first **A** from the first parent and the **A** from the second parent, which leads to the genotype **AA**.

(ii) The offspring gets the first **A** from the first parent and the **a** from the second parent, which leads to the genotype **Aa**.

(iii) The offspring gets the second **A** from the first parent and the **A** from the second parent leading to genotype **AA**.

(iv) The offspring gets the second **A** from the first parent and the **a** from the second parent leading to genotype **Aa**.

The probability of each of the above four cases is $\frac{1}{2} \times \frac{1}{2} = \frac{1}{4}$.

Thus, if the parent pairing is (**AA**,**Aa**), the offspring will always inherit an **A** gene from the first parent and either an **A** or an **a** (with equal probability) from the second parent. Thus the offspring is likely to inherit a genotype of **AA** or **Aa** with a probability of $1/2$ each. This implies that of all the offspring resulting from a parent pairing (**AA**,**Aa**), we expect 50% to have genotype **AA** and the other 50% to have genotype **Aa**, while there won't be any with genotype **aa**. All the offspring will therefore have red flowers.

If we list all possible parent pairings and compute the probabilities of the resulting offspring combinations we will arrive at Table 8.2.1, which we shall refer to as the *genotype probability matrix*. We have only partially filled this table. Let us explain its structure. The nine columns of data are arranged in three groups according to the first parent. For example, the three columns corresponding to the parent pairings (**AA**,**AA**), (**AA**,**Aa**) and (**AA**,**aa**) are grouped under **AA**. The rows represent the offspring outcomes for the different parent pairings.

For example, consider the column headed by **AA** under the **AA** group. This corresponds to the parent pairing (**AA**,**AA**). In this case all offspring will have genotype **AA** and there will not be any plants of genotypes **Aa** or **aa**. Therefore the numerical entries in this column are 1, 0, 0. The next column corresponds to the parent pairing (**AA**,**Aa**). As discussed above, in this case the offspring are equally likely to be **AA** or **Aa** and definitely can't be **aa**. So the column entries are 1/2, 1/2, 0. Further, if the parent pairing is (**AA**,**aa**), then each offspring will inherit an **A** gene from the first parent and an **a** from the second parent and will therefore have the genotype **Aa**. This is indicated by the entries 0, 1, 0 in the third column of data.

	AA			Aa			aa		
	AA	Aa	aa	AA	Aa	aa	AA	Aa	aa
AA	1	1/2	0						
Aa	0	1/2	1						
aa	0	0	0						

Table 8.2.1 Probabilities of offspring genotypes for different parent pairings

Investigation 8.9 Complete the genotype probabilities for all possible parent pairings in Table 8.2.1.

The matrix $M = \begin{bmatrix} 1 & 1/2 & 0 \\ 0 & 1/2 & 1 \\ 0 & 0 & 0 \end{bmatrix}$ (that is, columns 2, 3 and 4 of Table 8.2.1) is the fundamental genetic matrix which refers to all genotypes mating with **AA**. All of its elements are non-negative and the column sums are equal to 1. Such matrices are called *stochastic matrices*.

Investigation 8.10 Write the fundamental genetic matrices which refer to mating with the genotypes **Aa** and **aa**.

Let p, q and r denote the probabilities of an individual plant having genotype **AA**, **Aa** or **aa** respectively. Then $p + q + r = 1$, and $P = (p, q, r)$ is called the *probability vector* of the population. We wish to study how the genotype numbers will change in the next generation. Let us assume the simplest situation, where all members of the population are mated with plants of a single genotype, say **AA**. Recall that if the sample space of a random experiment is cut into three mutually disjoint parts X, Y, Z then the probability of an event E can be expressed in terms of conditional probabilities as follows.

$$P(E) = P(E|X)P(X) + P(E|Y)P(Y) + P(E|Z)P(Z)$$

Let X, Y, Z be the events that the second parent had genotype **AA**, **Aa**, **aa** respectively (the first parent always having genotype **AA**). Let E be the event that offspring had genotype **AA**. Then

$$P(E|X) = \text{Probability of } \mathbf{AA} \text{ offspring from } (\mathbf{AA},\mathbf{AA}) \text{ parents} = 1$$
$$P(E|Y) = \text{Probability of } \mathbf{AA} \text{ offspring from } (\mathbf{AA},\mathbf{Aa}) \text{ parents} = 1/2$$
$$P(E|Z) = \text{Probability of } \mathbf{AA} \text{ offspring from } (\mathbf{AA},\mathbf{aa}) \text{ parents} = 0$$

Therefore, the overall probability of an **AA** offspring is

$$P(E) = 1 \cdot p + \frac{1}{2} \cdot q + 0 \cdot r = p + \frac{q}{2}$$

We can similarly compute the probabilities of the other two genotypes in the next generation. This gives us the the following three computations.

$$\begin{aligned} 1 \cdot p + \frac{1}{2} \cdot q + 0 \cdot r &= p + \frac{q}{2} \\ 0 \cdot p + \frac{1}{2} \cdot q + 1 \cdot r &= \frac{q}{2} + r \\ 0 \cdot p + 0 \cdot q + 0 \cdot r &= 0 \end{aligned} \tag{8.1}$$

These equations can be expressed as a multiplication of matrices,

$$\begin{bmatrix} 1 & 1/2 & 0 \\ 0 & 1/2 & 1 \\ 0 & 0 & 0 \end{bmatrix} \begin{bmatrix} p \\ q \\ r \end{bmatrix} = \begin{bmatrix} p + \frac{q}{2} \\ \frac{q}{2} + r \\ 0 \end{bmatrix} \tag{8.2}$$

We see that the probability vector for the first generation offspring when all plants are fertilised with **AA**, will be obtained by multiplying the fundamental genetic matrix M with the initial probability vector P. In using matrices, we will write a vector in either row or column form, as convenient.

Investigation 8.11 Obtain the equations for the first generation offspring when all plants are fertilised with (a) **Aa** and (b) **aa**.

This lays the foundation for a mathematical model based on matrices, which can be used to predict the genotype distribution of the plant population after any number of generations under specific breeding programs, wherein each plant of the population is fertilised with a plant of a specific genotype.

The Matrix Model

For formulating the study of population vectors of many successive generations, let us assume a_n, b_n and c_n to be the fraction of plants of genotype **AA**, **Aa** and **aa** respectively in the n generation, where $n = 0, 1, 2, \ldots$ and $a_n + b_n + c_n = 1$ for every n^{th}. Here, a_0, b_0, c_0 denote the initial distribution of genotypes **AA**, **Aa** and **aa** respectively in the population. Let us denote the corresponding probability vectors by $x^{(n)}$. That is,

$$x^{(n)} = (a_n, b_n, c_n)$$

Consider the case when all the plants are fertilised with type **AA**. We have already established that multiplying a population vector by the fundamental genetic matrix gives the population vector of the next generation (see Eqs 8.1 and 8.2). In our current notation, this gives

$$\begin{aligned} a_n &= a_{n-1} + \frac{1}{2} b_{n-1} \\ b_n &= \frac{1}{2} b_{n-1} + c_{n-1} \\ c_n &= 0 \end{aligned} \tag{8.3}$$

or

$$x^{(n)} = Mx^{(n-1)}, \quad n = 1, 2, 3, \ldots \tag{8.4}$$

where

$$M = \begin{bmatrix} 1 & 1/2 & 0 \\ 0 & 1/2 & 1 \\ 0 & 0 & 0 \end{bmatrix}, \quad x^{(n)} = \begin{bmatrix} a_n \\ b_n \\ c_n \end{bmatrix}, \quad x^{(n-1)} = \begin{bmatrix} a_{n-1} \\ b_{n-1} \\ c_{n-1} \end{bmatrix}$$

Investigation 8.12 Using the equation $x^{(n)} = Mx^{(n-1)}$ verify that $x^{(n)} = M^n x^{(0)}$.

This matrix equation may be used to predict the distribution of genotypes in the nth generation if the initial distribution $x^{(0)}$ is known.

Let us suppose that in the initial generation, the distribution of the three genotypes are given by $a_0 = 0.3$, $b_0 = 0.5$, $c_0 = 0.2$ and that all plants are fertilised with plants of genotype **AA**. We shall now use the matrix equation $x^{(n)} = M^n x^{(0)}$ to compute the genotype distributions of the plant population in the next five generations.

$$\text{Thus } x^{(1)} = Mx^{(0)} = \begin{bmatrix} 1 & 1/2 & 0 \\ 0 & 1/2 & 1 \\ 0 & 0 & 0 \end{bmatrix} \begin{bmatrix} 0.3 \\ 0.5 \\ 0.2 \end{bmatrix} = \begin{bmatrix} 0.3 + 0.25 \\ 0.25 + 0.2 \\ 0 \end{bmatrix} = \begin{bmatrix} 0.55 \\ 0.45 \\ 0 \end{bmatrix}$$

Hence in the first generation, 55% of the plants will have genotype **AA** and the remaining 45% will have genotype **Aa**. There will be no plants of genotype **aa**.

We shall now use `Excel` to compute $x^{(2)}$ to $x^{(5)}$ (that is, the genotype distributions of the second, third, fourth and fifth generations).

Step 1: Enter matrix M in a 3×3 array (cells B3 to D6) and matrix $x^{(0)}$ as a 3×1 array (cells F3 to F6) as shown in Fig. 8.2.1. Select the 3×1 array (cells G3 to G6) for the product. Enter the `MMULT` command in cell G3, select the array B3 to D6 and the F3 to F6 successively, to multiply M and $x^{(0)}$. The product in the array G3 to G6 is $x^{(1)}$.

B	C	D	E	F	G	H	I	J	K
	M			x(0)	x(1)	x(2)	x(3)	x(4)	x(5)
1	0.5	0		0.3	=MMULT(B3:D5,F3:F5)				
0	0.5	1		0.5	MMULT(array1, array2)				
0	0	0		0.2					

Fig. 8.2.1

Step 2: To obtain $x^{(2)}$ use the `MMULT` command to multiply M with $x^{(1)}$ as done in step 1. The product in array H3 to H6 is $x^{(2)}$. In a similar manner, by multiplying M with $x^{(2)}$ obtain $x^{(3)}$ and so on. Figure 8.2.2 shows the genotype distributions from the first to the fifth generations.

B	C	D	E	F	G	H	I	J	K
	M			x(0)	x(1)	x(2)	x(3)	x(4)	x(5)
1	0.5	0		0.3	0.55	0.775	0.8875	0.94375	0.971875
0	0.5	1		0.5	0.45	0.225	0.1125	0.05625	0.028125
0	0	0		0.2	0	0	0	0	0

Fig. 8.2.2

Note that the genotype distribution of the 5th generation indicates that about 97% of the plants have genotype **AA** whereas only 3% of the plants have genotype **Aa**.

Investigation 8.13 Let us assume that we have a plant population with an initial distribution of genotypes given by $a_0 = 0.1$, $b_0 = 0.2$, $c_0 = 0.7$. Suppose all the plants are fertilised with plants of genotype **Aa**. What will be the genotype distributions after 5 generations? You may use `Excel` for computations.

A Mathematica Exploration

The following `Mathematica` code uses the equation $x^{(n)} = M^n x^{(0)}$ to compute the genotype distributions upto 20 generations.

```
M={{1, 0.5, 0},{0, 0.5, 1},{0, 0, 0}};
X0={{0.3},{0.5},{0.2}};
F[n_]:=MatrixPower[M,n].X0;
Table[F[n]//MatrixForm,{n,1,20}]
```

Figure 8.2.3 shows the first three and last three probability vectors from this calculation. We see that the **AA** fraction, given by the first entry of each probability vector, dominates the others and over time becomes all of the population.

$$\begin{pmatrix} 0.55 \\ 0.45 \\ 0. \end{pmatrix}, \begin{pmatrix} 0.775 \\ 0.225 \\ 0. \end{pmatrix}, \begin{pmatrix} 0.8875 \\ 0.1125 \\ 0. \end{pmatrix}, \ldots, \begin{pmatrix} 0.999997 \\ 3.43323\times10^{-6} \\ 0. \end{pmatrix}, \begin{pmatrix} 0.999998 \\ 1.71661\times10^{-6} \\ 0. \end{pmatrix}, \begin{pmatrix} 0.999999 \\ 8.58307\times10^{-7} \\ 0. \end{pmatrix}$$

Fig. 8.2.3 Portions of the `Mathematica` output describing the probability vectors for 20 generations.

The plots of the distributions of the three genotypes may be obtained by the additional line and the output is shown in Fig. 8.2.4.

```
ListPlot[Transpose[Table[Flatten[F[n]],{n,1,20}]],
 PlotTheme->"Monochrome",PlotLegends->{"AA","Aa","aa"}]
```

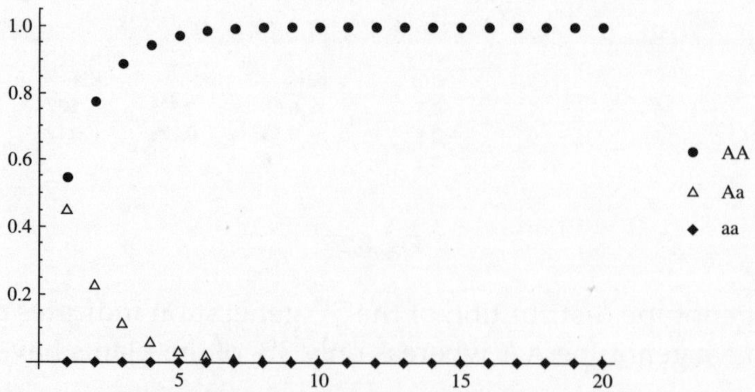

Fig. 8.2.4 `Mathematica` output showing that all plants will have genotype **AA** in the long run

Investigation 8.14 Modify the `Mathematica` code to obtain the genotype distributions of the plant population, given the different initial distributions $x^{(0)}$ below, if all the plants are fertilised with plants of genotype **Aa**. Also obtain plots of the population trends.

(a) $x^{(0)} = \begin{bmatrix} 0.5 \\ 0.3 \\ 0.2 \end{bmatrix}$ (b) $x^{(0)} = \begin{bmatrix} 0 \\ 0 \\ 1 \end{bmatrix}$

Further, what happens to the distribution of phenotypes, that is, the number of red and white flowers?

Eigenvalues and Eigenvectors

In the computations done so far we have used the matrix equation $x^{(n)} = M^n x^{(0)}$ to predict the genotype distribution of the plant population in any generation provided the initial distribution $x^{(0)}$ is known. However, one critical aspect is computing the nth power of the matrix M, which is difficult and time consuming without `Mathematica` or `Excel`. But can we find $M^n x^{(0)}$ without carrying out a large number of additions and multiplications? Matrix Algebra provides excellent tools known as *eigenvalues* and *eigenvectors* of a matrix. Their use involves more initial work, namely finding the roots of a polynomial equation, but after that the computations become very easy. Moreover, their use gives theoretical insight into the results of the computations.

We begin by identifying special vectors v for which the calculations $M^n v$ are easy. We say that a non-zero vector v is an *eigenvector* of M, with corresponding *eigenvalue* λ (a real number) if $Mv = \lambda v$. That is, the effect of M on v is merely to scale it by a real number.

Investigation 8.15 Suppose v is an eigenvector of M with eigenvalue λ. Show that $M^n v = \lambda^n v$.

How do we find eigenvectors for a given matrix? We first rearrange the equation $Mv = \lambda v$ as follows: $Mv = \lambda v \iff Mv - \lambda I v = 0 \iff (M - \lambda I)v = 0$, where I is the identity matrix.

Recall that if A is a square matrix, then $Av = 0$ has a non-zero solution v if and only if the determinant $|A|$ is zero. These two observations lead to the following result:

Theorem 8.2 The real number λ is an eigenvalue of the square matrix M if and only if $|M - \lambda I| = 0$.

The equation $|M - \lambda I| = 0$ is called the *characteristic equation* of M. When we calculate the determinant $|M - \lambda I|$ we obtain a polynomial in the variable λ. Since a polynomial may not have real roots, a matrix of real numbers may fail to have real eigenvalues.

Consider $M = \begin{bmatrix} 1 & 1/2 & 0 \\ 0 & 1/2 & 1 \\ 0 & 0 & 0 \end{bmatrix}$. To find its eigenvalues, we need to solve its characteristic equation $|M - \lambda I| = 0$ for λ:

$$0 = |M - \lambda I| = \left| \begin{bmatrix} 1 & 1/2 & 0 \\ 0 & 1/2 & 1 \\ 0 & 0 & 0 \end{bmatrix} - \lambda \begin{bmatrix} 1 & 0 & 0 \\ 0 & 1 & 0 \\ 0 & 0 & 1 \end{bmatrix} \right| = \begin{vmatrix} 1-\lambda & 1/2 & 0 \\ 0 & (1/2)-\lambda & 1 \\ 0 & 0 & -\lambda \end{vmatrix}$$

Investigation 8.16 Evaluate the above determinant and solve for λ to show that the eigenvalues of M are 1, 1/2 and 0.

Let us now find an eigenvector of M corresponding to the eigenvalue 1, by solving the equation $(M - I)v = 0$ for v.

$$\begin{bmatrix} 1-1 & 1/2 & 0 \\ 0 & 1/2-1 & 1 \\ 0 & 0 & 0-1 \end{bmatrix} \begin{bmatrix} v_1 \\ v_2 \\ v_3 \end{bmatrix} = \begin{bmatrix} 0 \\ 0 \\ 0 \end{bmatrix} \implies \begin{bmatrix} 0 & 1/2 & 0 \\ 0 & -1/2 & 1 \\ 0 & 0 & -1 \end{bmatrix} \begin{bmatrix} v_1 \\ v_2 \\ v_3 \end{bmatrix} = \begin{bmatrix} 0 \\ 0 \\ 0 \end{bmatrix}$$

This gives the following system of linear equations:

$$\begin{aligned} (1/2)v_2 &= 0 \\ -(1/2)v_2 + v_3 &= 0 \\ -v_3 &= 0 \end{aligned}$$

These equations yield $v_2 = v_3 = 0$ while v_1 can take any arbitrary value. Taking $v_1 = 1$ gives the eigenvector $v = \begin{bmatrix} 1 \\ 0 \\ 0 \end{bmatrix}$ corresponding to the eigenvalue $\lambda = 1$.

Investigation 8.17 Show that $\begin{bmatrix} -1 \\ 1 \\ 0 \end{bmatrix}$ and $\begin{bmatrix} 1 \\ -2 \\ 1 \end{bmatrix}$ are eigenvectors corresponding to the eigenvalues 1/2 and 0 (of the matrix M) respectively.

The following `Mathematica` code can be used to find the eigenvalues and eigenvectors of the matrix M.

```
M={{1, 1/2, 0},{0, 1/2, 1},{0, 0, 0}};
Eigenvalues[M]
Eigenvectors[M]
```

At this point, we have seen how to identify certain vectors v for which it is easy to compute $M^n v$. The next task is to use these special vectors as building blocks for others. For example, consider

$$\begin{bmatrix} 1 \\ -1 \\ 1 \end{bmatrix} = \begin{bmatrix} 1 \\ 0 \\ 0 \end{bmatrix} + \begin{bmatrix} -1 \\ 1 \\ 0 \end{bmatrix} + \begin{bmatrix} 1 \\ -2 \\ 1 \end{bmatrix}$$

We can compute $M^n \begin{bmatrix} 1 \\ -1 \\ 1 \end{bmatrix}$ as follows:

$$M^n \begin{bmatrix} 1 \\ -1 \\ 1 \end{bmatrix} = M^n \begin{bmatrix} 1 \\ 0 \\ 0 \end{bmatrix} + M^n \begin{bmatrix} -1 \\ 1 \\ 0 \end{bmatrix} + M^n \begin{bmatrix} 1 \\ -2 \\ 1 \end{bmatrix} = 1^n \begin{bmatrix} 1 \\ 0 \\ 0 \end{bmatrix} + \left(\frac{1}{2}\right)^n \begin{bmatrix} -1 \\ 1 \\ 0 \end{bmatrix} + 0^n \begin{bmatrix} 1 \\ -2 \\ 1 \end{bmatrix}$$

$$= \begin{bmatrix} 1-(1/2)^n \\ (1/2)^n \\ 0 \end{bmatrix} \to \begin{bmatrix} 1 \\ 0 \\ 0 \end{bmatrix} \quad \text{as } n \to \infty$$

Investigation 8.18 Let $x^{(0)}$ be the initial probability vector of a population with genotypes **AA**, **Aa** and **aa**. Let $M = \begin{bmatrix} 1 & 1/2 & 0 \\ 0 & 1/2 & 1 \\ 0 & 0 & 0 \end{bmatrix}$ be the fundamental genetic matrix corresponding to fertilisation with the **AA** genotype. Suppose that

$$x^{(0)} = \alpha \begin{bmatrix} 1 \\ 0 \\ 0 \end{bmatrix} + \beta \begin{bmatrix} -1 \\ 1 \\ 0 \end{bmatrix} + \gamma \begin{bmatrix} 1 \\ -2 \\ 1 \end{bmatrix}$$

Show that

(a) $\alpha = 1$

(b) $M^n x^{(0)} \to \begin{bmatrix} 1 \\ 0 \\ 0 \end{bmatrix}$ as $n \to \infty$

Investigation 8.19 Suppose that our initial population is a mix of plants with red flowers (genotypes **AA** and **Aa**) and plants with white flowers (genotype **aa**). What experiment can we do to estimate the **AA** and **Aa** fractions of the plants with red flowers?

Steady-state Genotype Distribution

In the analysis done so far, using the equation $x^{(n)} = M^n x^{(0)}$ for fertilisation by the **AA** genotype, we observed that the genotype distribution approaches $\begin{bmatrix} 1 \\ 0 \\ 0 \end{bmatrix}$. This final state is referred to as the *steady-state distribution*, and is reached irrespective of the initial distribution $x^{(0)} = \begin{bmatrix} a_0 \\ b_0 \\ c_0 \end{bmatrix}$. Perhaps this is not surprising, since repeated fertilisation with **AA** should lead to it swamping the others. What the mathematical modelling achieves is the description of just how quickly this will happen.

This implies that no matter what distribution of red and white flower plants we begin with, eventually all plants will have red flowers if the plants in each generation are fertilised with genotype **AA**.

When the repeated fertilisation is by **Aa** plants then there is a potential for all three types to persist. Our model can provide numerical predictions for the final break-up into genotypes.

Investigation 8.20 Find the steady-state distribution if all the plants are fertilised with genotype **Aa**. You may refer to your findings in Investigation 8.14.

Investigation 8.21 Find the steady-state distribution if all the plants are fertilised with genotype **aa**.

Investigation 8.22 Suppose each plant is fertilised by one with the same genotype (an extreme case of inbreeding). Use `Mathematica` to find the composition after 20 generations, given the following initial genotype distributions:

(a) $x^{(0)} = \begin{bmatrix} 0.1 \\ 0.8 \\ 0.1 \end{bmatrix}$ (b) $x^{(0)} = \begin{bmatrix} 0.8 \\ 0.1 \\ 0.1 \end{bmatrix}$ (c) $x^{(0)} = \begin{bmatrix} 0.4 \\ 0.4 \\ 0.2 \end{bmatrix}$

Investigation 8.23 Consider the situation of Investigation 8.22. Use eigenvalues and eigenvectors to find the dependence of the steady-state distribution on the initial genotype distribution.

Let us also consider what happens to a population in the 'wild' with random fertilisation among members of the same generation. We assume there are distinct generations, and fertilisation only occurs within members of the same generation to produce the next generation. By 'random' we mean that the probability of getting fertilised by, say, a member of the **AA** genotype is equal to the proportion of the population which is of that genotype. For example, if 30% of the population is **AA** then the probability of being fertilised by a plant with **AA** genotype 0.3. Let M_{AA}, M_{Aa} and M_{aa} be the fundamental genetic matrices for fertilising by the genotypes **AA**, **Aa** and **aa** respectively. If the probability vector for any generation is $\begin{bmatrix}\alpha\\ \beta\\ \gamma\end{bmatrix}$ then the probability vector for the next generation will be

$$\left(\alpha M_{AA} + \beta M_{Aa} + \gamma M_{aa}\right)\begin{bmatrix}\alpha\\ \beta\\ \gamma\end{bmatrix}$$

Investigation 8.24 Consider a population with random mating as described above. Show the following:

(a) Starting with any initial vector $x^{(0)}$, the probability vector for the next generation has the form $\begin{bmatrix}p^2\\ 2pq\\ q^2\end{bmatrix}$, with $p + q = 1$.

(b) The population vectors from $x^{(1)}$ onwards are constant. (This is called the *Hardy–Weinberg Equilibrium.*)

X-linked Inheritance

As we noted earlier, this particular vision of inheritance based on selecting one out of two genes from each parent is 'typical', and not universal. Another kind of inheritance is *X-linked inheritance* in which the male possesses one gene (**A** or **a**) and the female possesses two genes (**AA**, **Aa** or **aa**). A male offspring receives one of his mother's two genes with equal probability and the female offspring receives her father's gene and one of her mother's genes with equal probability. Thus, if the parent pairing is (**A**, **Aa**), a male offspring will receive **A** or **a** from his mother with probability 1/2 each, whereas a female will receive **A** from her father for sure and an **A** or **a** from her mother with probability 1/2 each. Note that this parent pairing cannot lead to a female offspring with genotype **aa**.

Investigation 8.25 There are six possible parent pairings as shown in the table. Complete the table by inserting the probabilities of the offspring outcomes. The entries of the column for the (**A**, **Aa**) parent pairing have been entered.

The Hardy–Weinberg Equilibrium of Investigation 8.24 is observed here as well, although the equilibrium occurs in the long-term and not immediately. Still, a population with random mating moves inexorably towards this equilibrium. We ask you to explore this phenomenon below.

		Parent pairings					
		(A,AA)	**(A,Aa)**	**(A,aa)**	**(a,AA)**	**(a,Aa)**	**(a,aa)**
Male offspring	**A**		1/2				
	a		1/2				
Female offspring	**AA**		1/2				
	Aa		1/2				
	aa		0				

Investigation 8.26 Consider a population of males and females, and an X-linked gene. Let p_A and p_a denote the fractions of the male population with genotype **A** and **a** respectively. Similarly, let q_{AA}, q_{Aa} and q_{aa} denote the fractions of the female population with the corresponding genotypes.

(a) Show that under random mating, a population vector $(p_A, p_a, q_{AA}, q_{Aa}, q_{aa})$ leads to the following composition of the next generation.

$$\left(q_{AA}+\frac{q_{Aa}}{2}, q_{aa}+\frac{q_{Aa}}{2}, p_A(q_{AA}+\frac{q_{Aa}}{2}), p_A(q_{aa}+\frac{q_{Aa}}{2})+p_a(q_{AA}+\frac{q_{Aa}}{2}), p_a(q_{aa}+\frac{q_{Aa}}{2})\right)$$

(b) Use `Mathematica` to plot how the genotype distribution changes over 10 generations, starting with an initial population vector $(1,0,0,1,0)$.

8.3 Age-structured Population Growth

In this section we shall explore an application of matrices to population growth. In particular we shall explore a model, which describes the growth of the female portion of a human or animal population which is divided into age classes. The model was proposed by Patrick H Leslie in the 1940s and is a popular model in population ecology.

We assume the female population to be divided into m age groups of equal time span. It is also assumed that birth and death rates are constant within each age group and over time. We will study this population at discrete times $t_0 = 0$, $t_1, t_2, \ldots$. For convenience we will take the gaps between these discrete times to be the same as the span of each age group. The advantage of this choice is that as time advances from t_n to t_{n+1}, each surviving member of an age group will definitely move to the next age group. Let $x_1^{(n)}, x_2^{(n)}, \ldots, x_m^{(n)}$ denote the respective populations of these m groups at time t_n. In particular, $x_i^{(0)}$ denotes the initial population of the ith group.

To fix these ideas better, we will build an example in parallel with the general approach. Consider the female portion of an animal population in which the highest age attained by a female is 24 years. We divide the members of the population into $m = 4$ age groups of 6 years each. Thus the age groups are 0 to 6 years, 6 to 12 years, 12 to 18 years and 18 to 24 years respectively. The discrete times at which we will model the population will similarly be $t_0 = 0$, $t_1 = 6, t_2 = 12, \ldots$. In this example, let us further assume that each group initially consisted of 100 females, so that $x_1^{(0)} = x_2^{(0)} = x_3^{(0)} = x_4^{(0)} = 100$.

Returning to our general model, let k_i, $i = 1, 2, 3, \ldots, n$ denote the average number of females who were born in any time interval (t_n, t_{n+1}) to each female in the ith age group and were alive at the time t_{n+1}. The k_i's are referred to as *fecundity ratios.*

Every newborn female of course begins life in the very first age group. Therefore, the number of females in the $i = 1$ class at time t_{n+1} equals the sum of the surviving numbers of those born to females in each class during the time interval (t_n, t_{n+1}). Thus,

$$x_1^{(n+1)} = k_1 x_1^{(n)} + k_2 x_2^{(n)} + \cdots + k_m x_m^{(n)} \tag{8.5}$$

In our running example, each time interval (t_n, t_{n+1}) comprises 6 years. Let the fecundity ratios of the four age classes be $k_1 = 0, k_2 = 4, k_3 = 3$ and $k_4 = 2$. Since each group initially had 100 females, the number of females born during the first 6 years that survive to the end of the 6th year is

$$x_1^{(1)} = 0 \cdot 100 + 4 \cdot 100 + 3 \cdot 100 + 2 \cdot 100 = 900$$

The number of females in the higher age groups $(i = 2, \ldots, m)$ at time t_{n+1} consists entirely of the survivors from the directly preceding groups. For example, the members of the $i = 2$ group will be the survivors from the $i = 1$ group. Let p_i denote the proportion of survivors in the ith class over any time interval (t_n, t_{n+1}). Then the populations in the different age groups at time t_{n+1} are given by

$$\begin{aligned} x_2^{(n+1)} &= p_1 x_1^{(n)} \\ x_3^{(n+1)} &= p_2 x_2^{(n)} \\ &\vdots \\ x_m^{(n+1)} &= p_{m-1} x_{m-1}^{(n)} \end{aligned} \tag{8.6}$$

In our example, let us suppose the survival proportions of each group are $p_1 = 1/2$, $p_2 = 1/3$ and $p_3 = 1/4$. There are no survivors from the final age group of 18–24 years. Then the age group populations of the second, third and fourth age groups at $t_1 = 6$ years are given by

$$x_2^{(1)} = 100/2 = 50, \quad x_3^{(1)} = 100/3 = 33\tfrac{1}{3}, \quad x_4^{(1)} = 100/4 = 25$$

Thus the model enables us to calculate the population of each age group after one time step.

Equations 8.5 and 8.6 may be conveniently written as a single matrix equation

$$\begin{bmatrix} x_1^{(n+1)} \\ x_2^{(n+1)} \\ x_3^{(n+1)} \\ \vdots \\ x_m^{(n+1)} \end{bmatrix} = \begin{bmatrix} k_1 & k_2 & \dots & k_{m-1} & k_m \\ p_1 & 0 & \dots & 0 & 0 \\ 0 & p_2 & \dots & 0 & 0 \\ \vdots & \vdots & \vdots & \vdots & \vdots \\ 0 & 0 & \dots & p_{m-1} & 0 \end{bmatrix} \begin{bmatrix} x_1^{(n)} \\ x_2^{(n)} \\ x_3^{(n)} \\ \vdots \\ x_m^{(n)} \end{bmatrix} \quad \text{or} \quad X^{(n+1)} = LX^{(n)} \tag{8.7}$$

where

$$L = \begin{bmatrix} k_1 & k_2 & \dots & k_{m-1} & k_m \\ p_1 & 0 & \dots & 0 & 0 \\ 0 & p_2 & \dots & 0 & 0 \\ \vdots & \vdots & \vdots & \vdots & \vdots \\ 0 & 0 & \dots & p_{m-1} & 0 \end{bmatrix}, \quad X^{(n+1)} = \begin{bmatrix} x_1^{(n+1)} \\ x_2^{(n+1)} \\ x_3^{(n+1)} \\ \vdots \\ x_m^{(n+1)} \end{bmatrix}, \quad X^{(n)} = \begin{bmatrix} x_1^{(n)} \\ x_2^{(n)} \\ x_3^{(n)} \\ \vdots \\ x_m^{(n)} \end{bmatrix}$$

The $n \times n$ matrix L is also known as the *Leslie matrix*. It has the following properties:

(i) All its elements are non-negative.

(ii) The elements p_i of its main sub-diagonal lie between 0 and 1.

(iii) The elements which are not in the first row or on the main sub-diagonal, are 0.

From (8.7) we have

$$\begin{aligned}
X^{(1)} &= LX^{(0)} \\
X^{(2)} &= LX^{(1)} = L^2X^{(0)} \\
X^{(3)} &= LX^{(2)} = L^3X^{(0)} \\
&\vdots \\
X^{(n)} &= LX^{(n-1)} = L^nX^{(0)}
\end{aligned} \tag{8.8}$$

The equation $X^{(n)} = L^nX^{(0)}$ helps us to compute the age distribution of the population at any stage if the initial distribution $X^{(0)}$ and the Leslie matrix L are known.

An example of the age-structured model

Consider again the (female) population in which the highest age attained by a female is 24 years. As mentioned earlier, the population is divided into four equal age classes of 6 years each. Thus the age classes are 0 to 6 years, 6 to 12 years, 12 to 18 years and 18 to 24 years respectively. We took the fecundity ratios of the four age classes to be $k_1 = 0$, $k_2 = 4$, $k_3 = 3$ and $k_4 = 2$. We let the survival proportions of the age groups be $p_1 = 1/2$, $p_2 = 1/3$ and $p_3 = 1/4$. Therefore the Leslie matrix for the population is given by

$$L = \begin{bmatrix} 0 & 4 & 3 & 2 \\ 1/2 & 0 & 0 & 0 \\ 0 & 1/3 & 0 & 0 \\ 0 & 0 & 1/4 & 0 \end{bmatrix}$$

If there are initially 100 females in each of the four age groups, we have

$$X^{(0)} = \begin{bmatrix} 100 \\ 100 \\ 100 \\ 100 \end{bmatrix}$$

After one time step, that is 6 years, the population is given by

$$X^{(1)} = LX^{(0)} = \begin{bmatrix} 0 & 4 & 3 & 2 \\ 1/2 & 0 & 0 & 0 \\ 0 & 1/3 & 0 & 0 \\ 0 & 0 & 1/4 & 0 \end{bmatrix} \begin{bmatrix} 100 \\ 100 \\ 100 \\ 100 \end{bmatrix} = \begin{bmatrix} 900 \\ 50 \\ 33\frac{1}{3} \\ 25 \end{bmatrix}$$

The matrix multiplication may be carried out in `Excel` or `Mathematica`, as discussed in the previous section and is shown in Fig. 8.3.1.

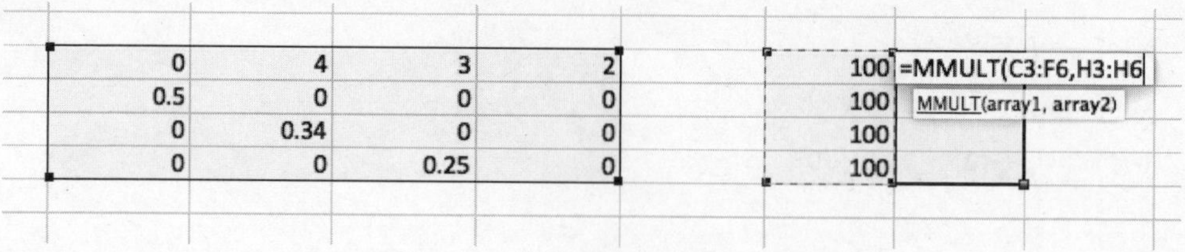

Fig. 8.3.1 Computing $X^{(1)}$ with `Excel`

Investigation 8.27 Find the distribution of the population after 30 years, using `Excel`. (Hint: A period of 30 years will include 5 time periods of 6 years each. The first time period is 0 to 6 years, the second is 6 to 12 years, and so on. The matrix $X^{(1)}$ gives us the population in the different age groups after 6 years, $X^{(2)}$ after 12 years, and so on.)

Let us also apply `Mathematica`. First, we compute and plot the population growth of each age group over 10 stages. The output is shown in Fig. 8.3.2.

```
leslie = {{0,4,3,2},{1/2,0,0,0},{0,1/3,0,0},{0,0,1/4,0}};
X0 = {{100},{100},{100},{100}};
F[n_] := MatrixPower[leslie,n].X0;
ListPlot[Transpose[Table[Flatten[F[n]], {n,1,10}]],
 Joined->True, PlotLegends->{"gp 1", "gp 2", "gp 3", "gp 4"}]
```

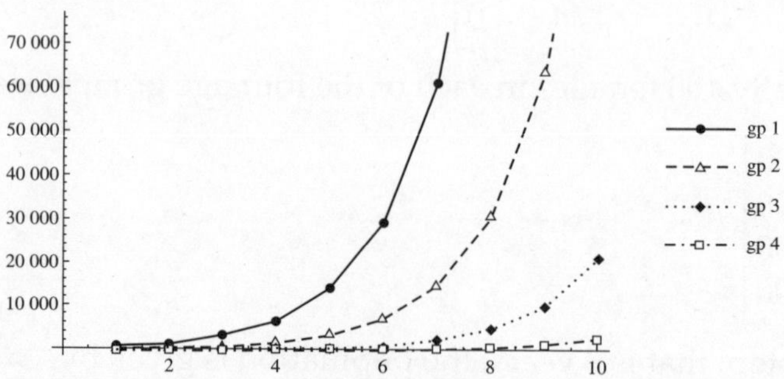

Fig. 8.3.2

The growth seems to be exponential for each group. To confirm this, we compute and plot the logarithms of the populations. The plot is shown in Fig. 8.3.3.

```
ListPlot[Transpose[Table[Flatten[Log[F[n]]], {n,1,10}]],
 Joined->True, PlotLegends->{"gp 1", "gp 2", "gp 3", "gp 4"}]
```

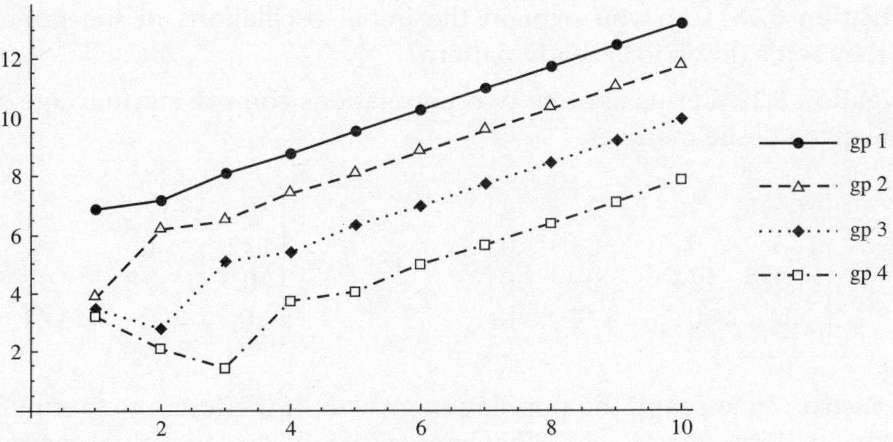

Fig. 8.3.3

The linear plots confirm the eventual exponential growth of each population. Finally, let us see how the composition of the population changes with time, over 30 years.

```
groups = N[Transpose[Table[Flatten[F[n]], {n,1,30}]]];
totals = Table[Sum[groups[[k]][[n]], {k,1,4}], {n,1,30}];
ListPlot[Table[groups[[k]]/totals, {k,1,4}],
  PlotLegends->{"gp 1","gp 2","gp 3", "gp 4"}]
```

We see in Fig. 8.3.4 that after some initial fluctuations, the age distribution settles down and moves towards a constant composition.

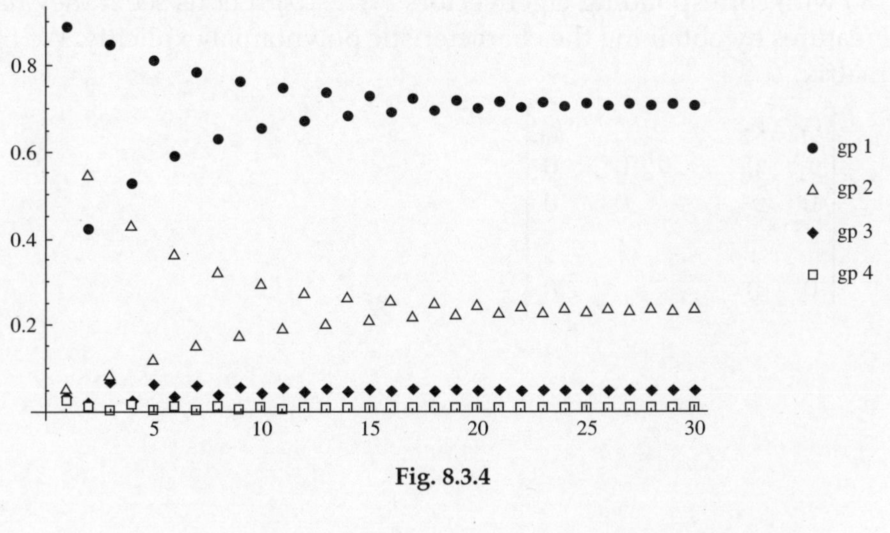

Fig. 8.3.4

Investigation 8.28 Can you explain the initial oscillations in the populations before they settle down to a steady pattern?

Investigation 8.29 Let us assume two populations comprising four age groups, with the given Leslie matrices.

$$\text{(a) } L = \begin{bmatrix} 0 & 2 & 3 & 4 \\ 1/3 & 0 & 0 & 0 \\ 0 & 1/4 & 0 & 0 \\ 0 & 0 & 1/4 & 0 \end{bmatrix}. \qquad \text{(b) } L = \begin{bmatrix} 0 & 1/2 & 2 & 1 \\ 1/2 & 0 & 0 & 0 \\ 0 & 1/3 & 0 & 0 \\ 0 & 0 & 1/4 & 0 \end{bmatrix}$$

Use `Mathematica` to graph the population growth of the four age groups in each case. What patterns do you see? What happens if you vary the initial compositions of the populations?

Limiting Behaviour

In this section, we look for theoretical support for the patterns we have seen through computations. We shall again use eigenvectors and eigenvalues. Let us recall that the eigenvalues λ of a square matrix A are the solutions of its characteristic equation $|A - \lambda I| = 0$. If A has size $d \times d$ then the expression $|A - \lambda I|$ is a polynomial of degree d in the variable λ. Hence the characteristic equation can have up to d distinct solutions, and these solutions would be a mix of real numbers and conjugate pairs of complex numbers.

We shall confine ourselves to the easiest case, when the characteristic equation of a Leslie matrix has d distinct real solutions. Then we get d distinct eigenvalues $\lambda_1, \ldots, \lambda_d$ with corresponding eigenvectors $v_1, \ldots, v_d$. Let us see if they have any special features by obtaining the characteristic polynomial explicitly. We first fix a Leslie matrix,

$$L = \begin{bmatrix} k_1 & k_2 & \ldots & k_{d-1} & k_d \\ p_1 & 0 & \ldots & 0 & 0 \\ 0 & p_2 & \ldots & 0 & 0 \\ \vdots & \vdots & \ddots & \vdots & \vdots \\ 0 & 0 & \ldots & p_{d-1} & 0 \end{bmatrix}$$

Next, we compute $|L - \lambda I|$ by repeatedly expanding down the first column.

$$|L-\lambda I| = \begin{vmatrix} k_1-\lambda & k_2 & \dots & k_{d-1} & k_d \\ p_1 & -\lambda & \dots & 0 & 0 \\ 0 & p_2 & \dots & 0 & 0 \\ \vdots & \vdots & \ddots & \vdots & \vdots \\ 0 & 0 & \dots & p_{d-1} & -\lambda \end{vmatrix}$$

$$= (k_1-\lambda)\begin{vmatrix} -\lambda & \dots & 0 & 0 \\ p_2 & \dots & 0 & 0 \\ \vdots & \ddots & \vdots & \vdots \\ 0 & \dots & p_{d-1} & -\lambda \end{vmatrix} - p_1 \begin{vmatrix} k_2 & \dots & k_{d-1} & k_d \\ p_2 & \dots & 0 & 0 \\ \vdots & \ddots & \vdots & \vdots \\ 0 & \dots & p_{d-1} & -\lambda \end{vmatrix}$$

$$= (k_1-\lambda)(-\lambda)^{d-1} - p_1 \begin{vmatrix} k_2 & \dots & k_{d-1} & k_d \\ p_2 & \dots & 0 & 0 \\ \vdots & \ddots & \vdots & \vdots \\ 0 & \dots & p_{d-1} & -\lambda \end{vmatrix}$$

We repeat this calculation with the last determinant, and eventually reach the characteristic equation

$$(-\lambda)^d + k_1(-\lambda)^{d-1} - p_1k_2(-\lambda)^{d-2} + p_1p_2k_3(-\lambda)^{d-3} + \dots + (-1)^{d-1}p_1\cdots p_{d-1}k_d = 0$$

Multiply through by $(-1)^d$ to get

$$\lambda^d - k_1\lambda^{d-1} - p_1k_2\lambda^{d-2} - p_1p_2k_3\lambda^{d-3} - \dots - p_1\cdots p_{d-1}k_d = 0$$

Divide through by λ^d and rearrange to get

$$1 = \frac{k_1}{\lambda} + \frac{p_1k_2}{\lambda^2} + \dots + \frac{p_1\cdots p_{d-1}k_d}{\lambda^d}$$

The right-hand side is large, certainly above 1, when λ is just to the right of zero. On the other hand, as λ increases, the right-hand side approaches zero. Hence it must take the value 1 somewhere. Moreover, it is a strictly decreasing function, and so can only be 1 at exactly one place. This shows that the characteristic equation has a unique positive solution, and the other $d-1$ solutions must be non-negative.

Investigation 8.30 Let λ_1 be the positive eigenvalue of the Leslie matrix L. Show that the corresponding eigenvector v_1 can be chosen to have the form

$$\begin{bmatrix} 1 \\ p_1/\lambda_1 \\ p_1p_2/\lambda_1^2 \\ \vdots \\ p_1\cdots p_{d-1}/\lambda_1^{d-1} \end{bmatrix}$$

Given an initial population vector $X^{(0)}$, we first express it as a combination of the d eigenvectors of L,

$$X^{(0)} = c_1v_1 + \cdots + c_dv_d$$

The later population vectors are given by

$$X^{(n)} = L^nX^{(0)} = c_1\lambda_1^nv_1 + \cdots + c_n\lambda_d^nv_d$$

The term corresponding to the eigenvalue with the largest magnitude will dominate as n increases. If that eigenvalue is negative, then the population will oscillate between positive and negative values, which is impossible. Hence the unique positive eigenvalue λ_1 must also have the largest magnitude. We call it the *dominant* eigenvalue.

Thus, for large n, we have

$$X^{(n)} \approx c_1\lambda_1^nv_1$$

Therefore, in the long run, the population vectors are scalar multiples of the eigenvector v_1.

Although we have established these patterns in a special case, they turn out to hold more generally—even when there are complex or repeated eigenvalues.

For example, consider the Leslie matrix we have already been working with,

$$L = \begin{bmatrix} 0 & 4 & 3 & 2 \\ 1/2 & 0 & 0 & 0 \\ 0 & 1/3 & 0 & 0 \\ 0 & 0 & 1/4 & 0 \end{bmatrix}$$

We compute its eigenvalues and their absolute values using `Mathematica`.

```
L=({{0,4,3,2},{1/2,0,0,0},{0,1/3,0,0},{0,0,1/4,0}});
eigen=N[Eigenvalues[L]]
Abs[eigen]
```

The output is given below.

```
{1.53646, -1.28931, -0.123576+0.163694 i, -0.123576-0.163694 i}
{1.53646, 1.28931, 0.205102, 0.205102}
```

We see that $\lambda_1 = 1.53646$ is the only positive eigenvalue and it also has the greatest absolute value. We compute the corresponding eigenvector and scale it so that the entries sum to 1.

$$\begin{bmatrix} 1 \\ p_1/\lambda_1 \\ p_1p_2/\lambda_1^2 \\ p_1p_2p_3/\lambda_1^3 \end{bmatrix} = \begin{bmatrix} 1 \\ 0.3254 \\ 0.0706 \\ 0.0115 \end{bmatrix} \rightarrow \begin{bmatrix} 0.71 \\ 0.23 \\ 0.05 \\ 0.01 \end{bmatrix}$$

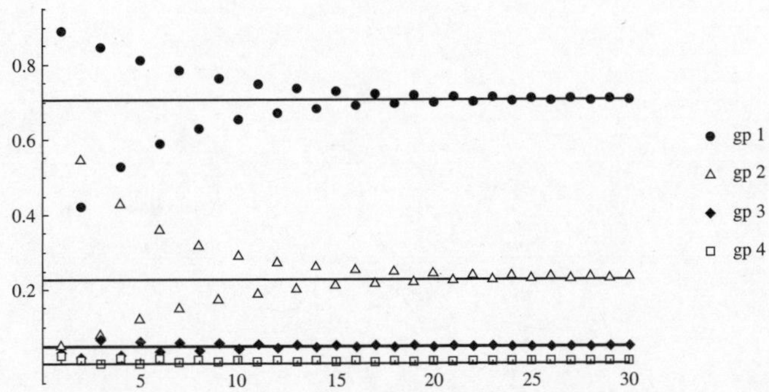

Fig. 8.3.5 The age distributions of Fig. 8.3.4 converge to (0.71, 0.23, 0.05, 0.01)

Investigation 8.31 Explore the limiting behaviour of the population with the Leslie matrix

$$\begin{bmatrix} 1 & 2 & 2 & 2 \\ 1/2 & 0 & 0 & 0 \\ 0 & 1/3 & 0 & 0 \\ 0 & 0 & 1/4 & 0 \end{bmatrix}$$

Compute the dominant eigenvalue and the equilibrium age distribution. Match these with direct computations of the age distribution after 20 time steps.

FURTHER EXPLORATIONS

Matrices have wide uses in many areas of mathematics and its applications. In this chapter we have given you a glimpse of some interesting applications of matrices to the topics of ranking, genetics and age-structured population growth. The reader may refer to the book by Anton and Rorres [2] to revisit different concepts and topics of Linear Algebra, while Kapur [24] provides extensive and detailed analyses of a rich variety of mathematical models related to multispecies population dynamics, optimal exploitation of resources, genetic improvement and inbreeding, and pharmacokinetics.

The articles by Joshi [21, 22, 23] provide an accessible treatment of the principles of genetics and how algebra can help compute their consequences. Leslie's 1945 article [27] is the original reference for applying matrices to understand age-structured populations. A related topic is the study of populations with distinct stages, such as those of the metamorphosis of a butterfly. For this, see the article by Lefkovitch [26]. Smith and Keyfitz [67] have collected a wide variety of articles related to the mathematical treatment of aspects of populations.

A

Introduction to Software

A.1 EXCEL

`Microsoft Excel` is a spreadsheet program, which is very handy for performing numerical calculations, for representing data using charts and graphs, and for simulating experiments. In most of the chapters of this book, `Excel` is one of the primary vehicles for such explorations. The steps of going through the explorations using `Excel` have been included within the chapters, as and when necessary. In this section, we provide a basic introduction to help you get started.

Identifying and selecting cells

When you click on the `Excel` icon, `Workbook1` opens up. A workbook comprises worksheets and each worksheet has a grid of cells in which the columns are labelled with letters A, B, C, D and so on, while the rows are numbered as 1,2,3,... This helps us to identify a particular cell using a cell address, that is, a letter followed by a number. Thus `D7` is the cell in row 7 and column D. You can select a cell simply by clicking on it. Once a cell has been selected a blue border will appear on the boundary of the cell as shown in Fig. A.1.1.

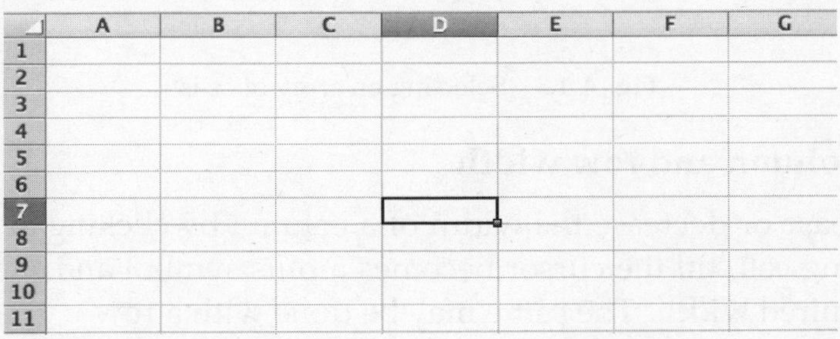

Fig. A.1.1 Selecting a cell in an Excel workbook

If you want to select a group of cells, say from `C1` to `C10`, you need to first select cell `C1`, press `shift` and then select cell `C10`. However if you wish to select the entire column C, you need to click on the column heading C. Similarly, a group of cells within a row can also be selected. If you want to select the entire row, say row 8, then you will have to click on the row heading on number 8 (Fig. A.1.2).

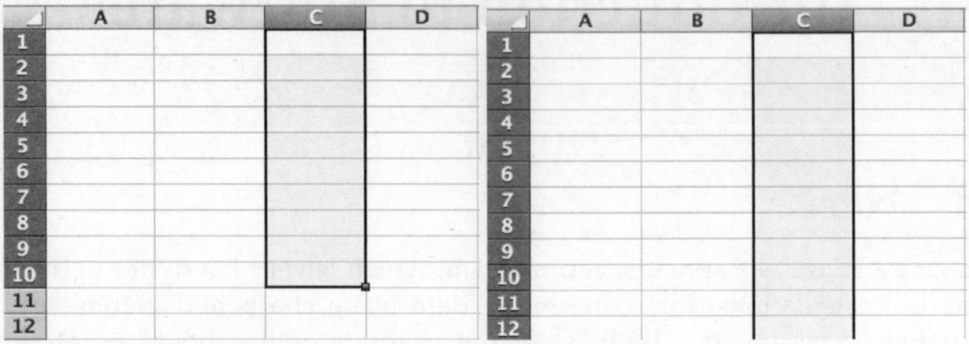

Fig. A.1.2 Selecting a column or a few cells in a column

You can also select an array of cells. This is done by clicking on the top left corner cell of the required array, pressing `shift` and then selecting the bottom right corner cell. For example, to select a 4 × 4 array, starting with `C4`, we first click on `C4`, press `shift` and then click on `F7` (Fig. A.1.3).

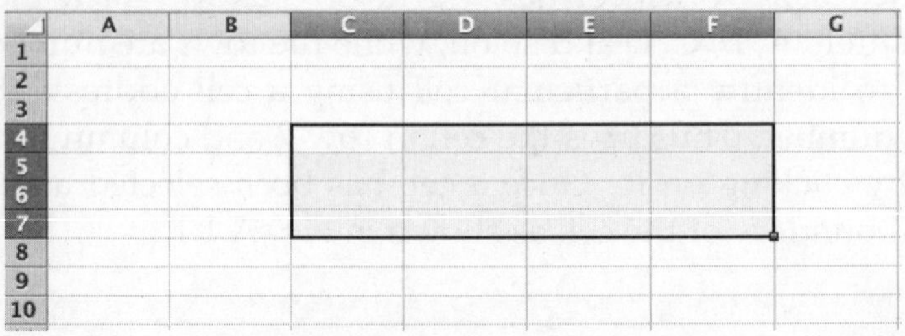

Fig. A.1.3 Selecting an array of cells

Modifying column and row width

You may increase or decrease the width of a column by clicking at the end of the column heading cell, till the cursor becomes a plus symbol and then dragging to obtain the required width. The same may be done with a row.

To create a column of numbers

If you wish to create a column of numbers from 1 to 100, you may do one of the following:

Method 1: Enter 1 in cell `A1` and the formula `=A1+1` in cell `A2`. Now bring the cursor to the corner of cell `A2` (till it resembles a small plus symbol) and then drag till cell `A100`.

Method 2: Enter 1 in cell `A1` and the formula `=A1+1` in cell `A2`. Then select cell `A2`, click on `Fill` (from the toolbar) and select `Series` from the options. In the `series` window, select `columns` (in the `Series in` option) and enter `Step value` as 1 and `Stop value` as 100. Finally click on `OK` to obtain the column of numbers. Figure A.1.4 shows the `Series` option window.

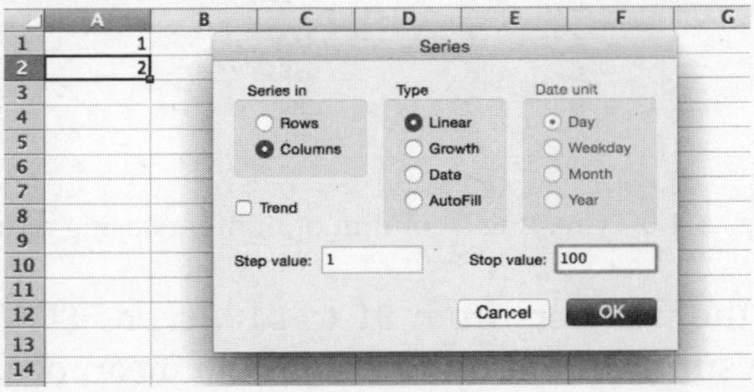

Fig. A.1.4 Using the Fill option to create a column of numbers

Basic numerical calculations and graphing

To perform a simple calculation such as multiplying two numbers, 875 and 423, click on any cell, type `=875*423` and press `Enter`. The answer will appear in the same cell. If you now select the cell, the formula entered in that cell will show up in the formula bar as shown in Fig. A.1.5. (Note that while entering any rule or formula in a cell, we type the = sign first. After typing the rule we have to press the `Enter` key.)

To compute the power of a number we use the ^ symbol appearing above the number 6 on the computer keyboard. For example, to calculate 23^4 we need to type `=23^4` in a cell. Similarly to compute the square root of a number we use the `SQRT` command. Thus to compute the square root of 637 we need to type `=SQRT(637)` in a cell.

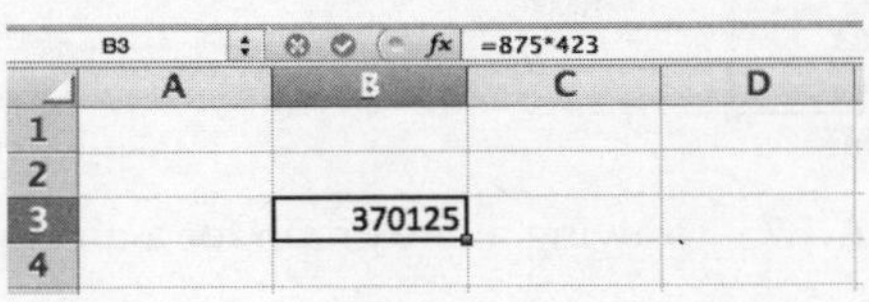

Fig. A.1.5 Performing calculations in `Excel`

If you want to create a column comprising, say, the first ten powers of 3, it is helpful to begin by creating a column of numbers from 1 to 10 from `A1` to `A10`. After this, type the formula `=3^A1` in cell `B1`, and double click on the corner of cell `B1`. This will create a column of the powers of 3. The output is shown in Fig. A.1.6.

	A	B
1	1	3
2	2	9
3	3	27
4	4	81
5	5	243
6	6	729
7	7	2187
8	8	6561
9	9	19683
10	10	59049

Fig. A.1.6 Creating a column of numbers using a rule

If you want to add the numbers from `B1` to `B10`, type `=SUM(B1:B10)` (in the cell where you want the sum). Similarly to obtain the mean of the numbers in `B1` to `B10`, type `=AVERAGE(B1:B10)`.

Also if you wish to copy the data (that is, powers of 3) from column B to column E, proceed as follows: Select the cells `B1` to `B10`, go to `Edit` (on the toolbar) select `Copy`, then go to column E, select cells `E1` to `E10`, right click and select `Paste Special` followed by `Values`.

In order to represent the data in column B graphically, select the cells `B1` to `B10`, go to `Charts` in the toolbar and select a `line graph`. The graph is shown in Fig. A.1.7.

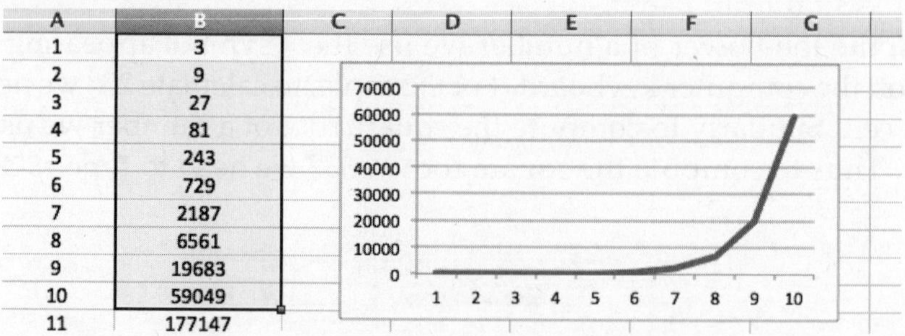

Fig. A.1.7 Drawing a line graph for a given data

`Excel` is a very powerful tool and has many useful functions. To learn more details about its basic features, you may refer to [25].

A.2 MATHEMATICA

Mathematica is a Computer Algebra System (CAS). It is capable of symbolic manipulation, so that it can do exact mathematics and not just numerical approximations. While it is a commercial software, there is an option to access it online on the Wolfram Cloud website (https://www.wolframcloud.com/). You can create a free account on Wolfram Cloud and save your files there. There are restrictions on the complexity of computations you can carry out with the free version, but these should not trouble a beginner. All the sample Mathematica code in this book has run without trouble in a free account.

Once your account is created, click on 'New Notebook' to create a file in which you can write and execute your commands. For example, you may begin by typing 1+2. To carry out this operation, simultaneously press the SHIFT and ENTER keys. A 3 will appear in the next line. At this point, the notebook will show the following.

```
In[1]:=1+2
Out[1]:=3
```

That is, the first input was 1+2 and the first output was 3. One can use Out[1] at any later stage to call on and use this particular calculation. For example, we can multiply this output by 4.

```
In[2]:=Out[1]*4
Out[2]:=12
```

The most recent output can also be used by calling on %. For example, we can calculate 12^2 as follows.

```
In[3]:=%^2
Out[3]:=144
```

Of course we could do several operations in one go, such as $((1+2)\times 4)^2/3$.

```
In[4]:=((1+2)*4)^2/3
Out[4]:=48
```

This example also illustrates that Mathematica knows the rules about brackets and the order in which arithmetic operations should be carried out. Further, the standard useful functions like square root, exponential, logarithms and the trigonometric functions are available with their usual names. The convention is that the name of any function will start with a capital letter, and the input will be enclosed within square brackets. `Mathematica` is case sensitive. The constants π and e are also available, as `Pi` and `E` respectively.

```
In[5]:=Cos[3*Pi]
Out[5]:=-1
In[6]:=Log[E^2]
Out[6]:=2
```

We can define and use our own functions.

```
In[7]:=Myfn[x_]:=Sqrt[7+x^2]
In[8]:=Myfn[3]
Out[8]:=4
```

Derivatives can be obtained using the f' notation.

```
In[9]:=Cos'[x]
Out[9]:=-Sin[x]
```

We can compute definite and indefinite integrals.

```
In[10]:=Integrate[Cos[x],x]
Out[10]:=Sin[x]
In[11]:=Integrate[Cos[x],{x,0,Pi/2}]
Out[11]:=1
```

Lists are placed within braces. For example, in the above definite integral, `{x,0,Pi/2}` is a list whose first entry states the variable of integration, while the second and third entries give the range of integration. The `Table` command can be used to create lists according to a certain pattern. A function can be applied to all the entries of a list, by just taking that list as input with the usual notation. In the code below, the command `N` is used to force an output in floating-point notation.

```
In[12]:=data=Table[n^2+1,{n,0,4}]
Out[12]:={1,2,5,10,17}
In[13]:=Sqrt[N[data]]
Out[13]:={1.,1.41421,2.23607,3.16228,4.12311}
```

The output of a particular command can be stopped from printing by closing the command with a semi-colon. This is especially useful when we wish to execute several lines in one go and only wish to see the final result. For example, here is a modification of the previous calculation.

```
In[14]:=data=Table[n^2+1,{n,0,4}];
        Sqrt[N[data]]
Out[14]:={1.,1.41421,2.23607,3.16228,4.12311}
```

In the code given above one presses ENTER after typing the first line to go to the next line without executing the first one. Once both lines have been typed, we press SHIFT and ENTER to execute them.

Individual entries of a list can be accessed by giving their position.

```
In[15]:=data[[3]]
Out[15]:=5
```

The `If` command lets us execute a choice of commands depending on whether some condition is satisfied. Within `If`, we first enter the condition to be checked, then what is to be done if the condition is met, and finally what is to be done if the condition is not met.

```
In[16]:=If[data[[3]]<=3,data[[3]],0]
Out[16]:=0
```

Here we first check if the third entry of the list `data` is less than or equal to 3. If it is, we will type its value. If it is not, we will type 0. Since the value is 5, the final output is 0.

The `While` command lets us repeat some action as long as a condition is met. It has two inputs: the condition to be met and the actions to be repeated.

```
In[17]:=n=1;While[data[[n]]<=5,Print[data[[n]]];n++]
        1
        2
        3
```

We start by setting $n = 1$ and then check if `data[[1]]` is less than or equal to 5. It is, so we print its value. The `n++` command then increases the value of n by one, and we repeat the process with $n = 2$.

This very basic introduction to `Mathematica` should suffice to get you started. The detailed pieces of code provided in the text will enable you to gain more expertise as you work through the book. For more structured learning, you can read [73] or visit the website [65].

A.3 GeoGebra

`GeoGebra` belongs to a class of software known as Dynamic Geometry Software (DGS). It enables the user to explore mathematical concepts and figures, graphically, symbolically and numerically. To download the software go to `https://www.geogebra.org/download` and select the current version of `Geogebra Classic`. The descriptions and screenshots below correspond to version 5.

As soon as a new document is opened in `GeoGebra`, it presents a `Graphics view` (with the axes and a grid in the background), an `Algebra view`, the `Toolbar` on top and an `Input Bar` at the bottom of the window as shown in Fig. A.3.1.

To remove the axes and the grid from the `Graphics view`, we can click on the tiny arrow on the left of the Graphics heading, and then choose the `Show or hide axes` and `Show or hide grid` options.

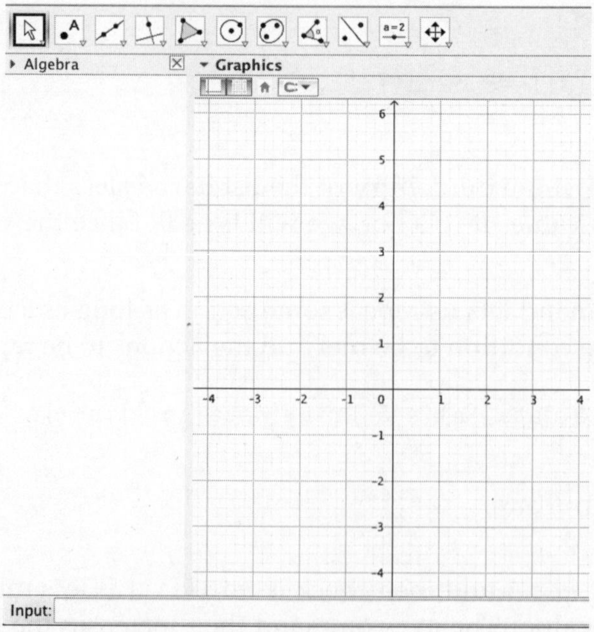

Fig. A.3.1 The **Graphics** view, **Algebra** view, the **Toolbar** and **Input** bar in a new `GeoGebra` document

Other views can be accessed using the `View` option on the toolbar. For example, by selecting `Spreadsheet` or `CAS` from the View option one can bring in a `spreadsheet` and a `CAS` window as well. The `Graphics` view, `Algebra` view, `Spreadsheet` view and `CAS` window can be used simultaneously to explore properties of a figure geometrically, symbolically as well as numerically (Fig. A.3.2).

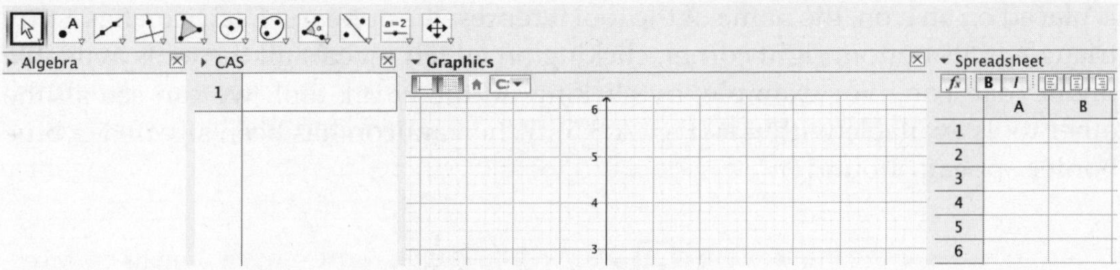

Fig. A.3.2 The **Algebra** view, **Graphics** view, **Spreadsheet** view and **CAS** window are simultaneously accessible in a `GeoGebra` document

Note that, once selected, these views will be ticked in the `View` option as shown in Fig. A.3.3.

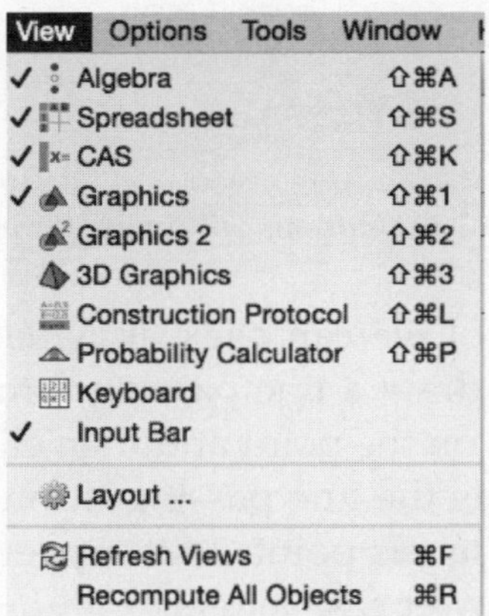

Fig. A.3.3 The **View** option allows the user to select different views

The GeoGebra Toolbar

The `GeoGebra` window has a Toolbar on top (Fig. A.3.4), which offers many options for drawing and manipulating figures. The Toolbar consists of 11 icons which represent the following tools (in sequence from the left): `Move`, `Point`, `Line`, `Perpendicular Line`, `Polygon`, `Circle with center through point`, `Ellipse`, `Angle`, `Reflect about line`, `Slider` and `Move graphics view`.

Fig. A.3.4 The GeoGebra toolbar

Each of these icons leads to a set of tools of a particular functionality. As the cursor is placed on an icon, the name of the tool it represents appears. Each icon has a tiny triangle in its bottom right corner, clicking on which reveals all the tools available under that icon. For example, by clicking on the `Point` tool, we can see all the other tools available within it (Fig. A.3.5). When an icon has been selected, a blue border appears around it.

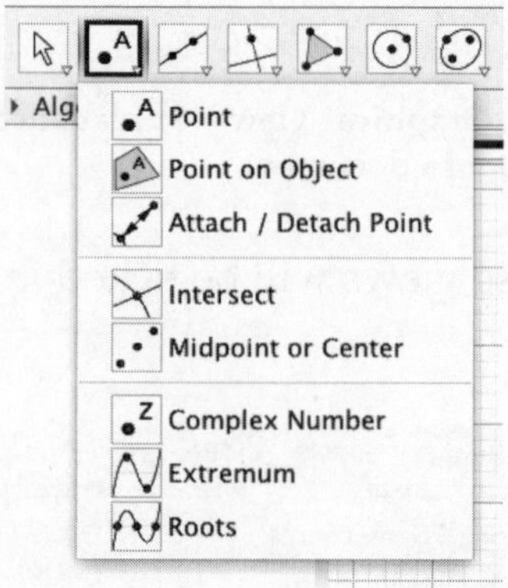

Fig. A.3.5 Options within the **Point** tool

After selecting the `Point` tool we can click anywhere on the `Graphics` view to obtain a point. Similarly to draw a line passing through a point, we select `Line` from `Line` tool options, click on the point and then click at another position in the `Graphics` view. By doing this the line passing through two given points will be drawn. Note that `GeoGebra` labels points with upper case letters while lines and segments are labelled with lower case letters.

Drawing Figures in GeoGebra

Let us draw a quadrilateral in the `Graphics` view using the `Polygon` tool of the `GeoGebra` toolbar as shown in Fig. A.3.6. To do this, select the `Polygon` icon from the tool bar and click anywhere on the `Graphics` view. `GeoGebra` will mark the first point as A. Proceed to draw a quadrilateral by marking the remaining three vertices. This is done by clicking on three other locations, which will be labelled as points B, C and D. To complete the quadrilateral, return to point A and click on it again. This completes the drawing of the quadrilateral ABCD. Note that `GeoGebra` immediately displays the coordinates of the vertices A,B,C,D the lengths of the sides a,b,c,d and the area of the quadrilateral (displayed as q1) in the `Algebra` view. This reiterates the point that a quadrilateral is a two-dimensional polygonal region in the plane.

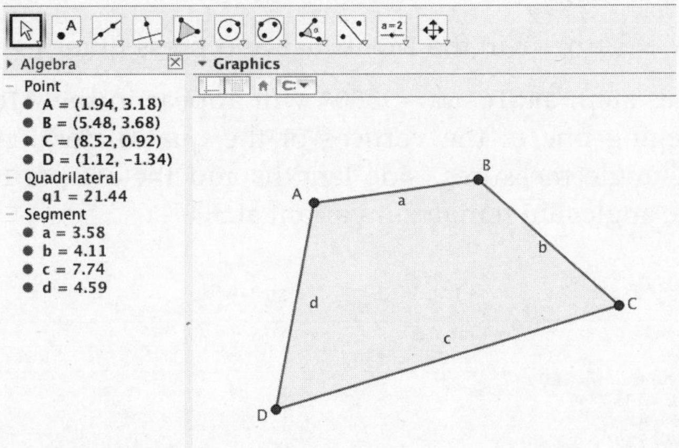

Fig. A.3.6 A quadrilateral drawn using the **Polygon** tool whose parts are named and described in the **Algebra** view

Selecting one of the vertices, say A, and dragging it will lead to varying experiences of the quadrilateral while its measurements will change dynamically in the `Algebra` view. Note that in order to select any part of a figure or to drag any point the `Move` tool (first icon of the toolbar) has to be selected. Figure A.3.7 shows another quadrilateral (resembling a parallelogram) obtained by dragging the vertices of the quadrilateral ABCD, one at a time.

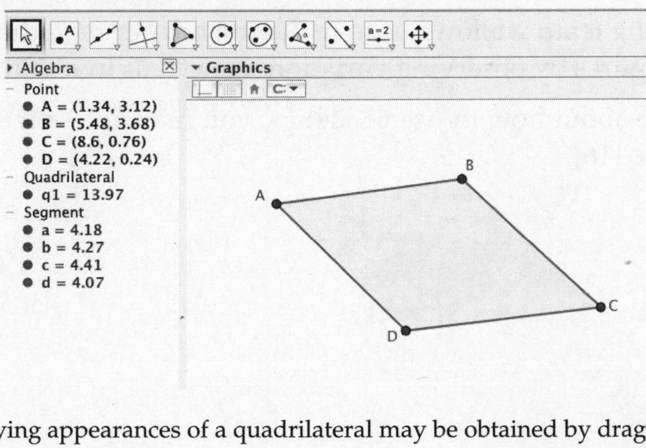

Fig. A.3.7 Varying appearances of a quadrilateral may be obtained by dragging the vertices

One can measure the interior angles of the quadrilateral ABCD by clicking on the `Angle` tool (on the toolbar) and then marking each angle. To mark an angle, say angle ADC, click on the points C, D and A (in clockwise order). The measures of the angles as well as the angles themselves will be marked inside the quadrilateral. Now type the expression

anglesum = $\alpha + \beta + \gamma + \delta$ in the `Input bar` and press Enter.

As a result of this step, `anglesum = 360°` will appear in the **Algebra** view (Fig. A.3.8). By dragging one of the vertices of the quadrilateral ABCD, you may observe that the angle measures, side lengths and the area of the quadrilateral vary whereas the anglesum remains invariant at 360°.

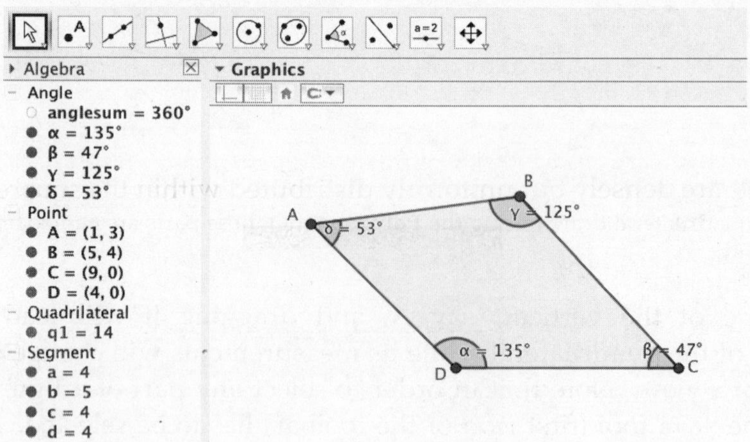

Fig. A.3.8 `GeoGebra` measures the sum of angles as 360° in the **Algebra** view

Thus, the dragging feature allows us to make conjectures regarding the properties of geometrical figures by observing variations as well as invariances.

To learn more about how to use `GeoGebra`, you may access the tutorials on the `GeoGebra` website [16].

B

Investigations: Hints and Results

B.1 CHAPTER 1

Investigation 1.1 The coordinates of the corners of the square are $(-1,1), (1,1)$, $(1,-1)$ and $(-1,-1)$.

Investigation 1.2 When we increase the number of simulated points to 10000, we find that they are densely but uniformly distributed within the square.

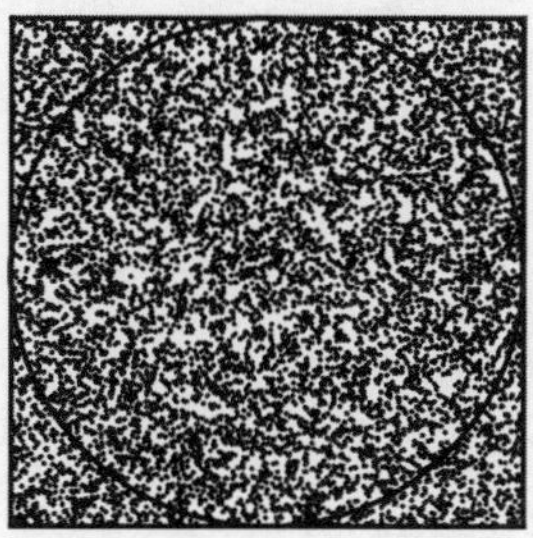

Fig. B.1.1 `Mathematica` output for 10000 simulated points

Investigation 1.3

Number of simulations	Number of hits (in 1000 throws)
1	785
2	782
3	772
4	789
5	784
6	784
7	805
8	780
9	803
10	788
Total (in 10000 throws)	7872

Fig. B.1.2 A sample set of 10 simulations. Your simulations should produce different numbers!

Investigation 1.6

(a) $(\theta, x) = (0,0)$

(b) $(\theta, x) = (0, x)$ with $0 < x \leq l/2$

(c) $(\theta, x) = (\pi/2, 0)$

(d) $(\theta, x) = (\pi/2, l/2)$

(e) $(\theta, x) = (\pi/3, x)$ with $0 \leq x \leq l/2$

Investigation 1.7 When the needle lands exactly between two lines (either perpendicular or parallel to the lines), the distance x of the centre from either line is $l/2$. When the needle lands in such a way that its centre is on a line then $x = 0$. Thus $0 \leq x \leq l/2$. If the needle lands exactly between two lines (parallel to the lines) or lands on a line, then the angle made by it with the horizontal is 0. Thus $\theta = 0$. If it lands perpendicular to the given lines then $\theta = \pi/2$. Thus $0 \leq \theta \leq \pi/2$.

Investigation 1.8 The area under the graph of $x = (l/2)\sin\theta$ is given by

$$\int_0^{\pi/2} x\, d\theta = \frac{l}{2}\int_0^{\pi/2} \sin\theta\, d\theta = \frac{l}{2}(-\cos\theta)\Big|_0^{\pi/2} = \frac{l}{2}[-\cos\frac{\pi}{2} + \cos 0] = \frac{l}{2}$$

Since the area of the rectangle of all possible needle throws is $(l/2) \times (\pi/2) = l\pi/4$, we have $p = \dfrac{l/2}{l\pi/4} = \dfrac{2}{\pi}$.

Investigation 1.11 After running the `Mathematica` code 10 times we will obtain 10,000 throws. We need to total the number of hits and use the formula $\frac{2}{\pi} \approx \frac{\text{number of hits}}{10000}$ to estimate the value of π. In general, this should lead to a better approximation of π than the one obtained from 1000 throws.

Investigation 1.13 When the length k of the needle is less than l, the distance between the lines, then the hits will be represented by the area under the graph $x = \frac{k}{2}\sin\theta$, which equals $\frac{k}{2}$ (*less than* l). However, the area of the rectangle representing all possible needle throws is still $\frac{l}{2} \times \frac{\pi}{2} = \frac{l\pi}{4}$. Thus the probability of a hit will be given by $p = \frac{k/2}{l\pi/4} = \frac{2k}{l\pi}$. For example, if $l = 1$ and $k = 2/3$ then $p = 4/3\pi = 0.424$. Hence the probability of a randomly thrown needle of length 2/3 units hitting a line is 0.424. Thus, smaller the value of k, the lesser will be the chance of the needle hitting a line.

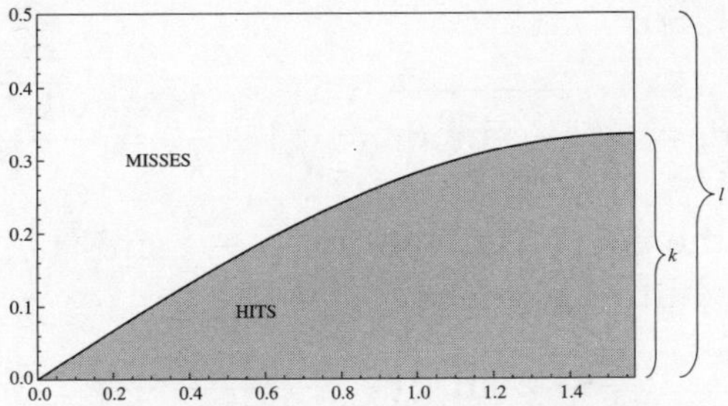

Fig. B.1.3 The $k < l$ case

Investigation 1.14 When the length k of the needle is greater than l, the distance between the lines, then the graph $x = \frac{k}{2} \sin\theta$ intersects the line $x = l/2$. The value of θ for which this happens will be obtained as follows:

$$\frac{l}{2} = \frac{k}{2}\sin\theta, \text{ which implies } \sin\theta = \frac{l}{k} \text{ or } \theta = \sin^{-1}(l/k)$$

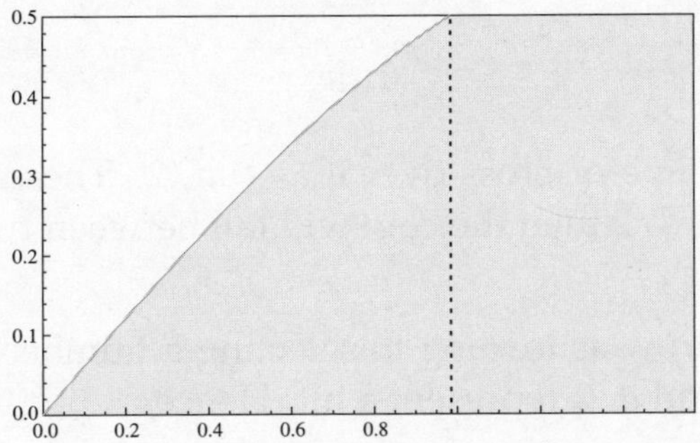

The $k < l$ case.

Fig. B.1.4 The $k > l$ case

The shaded area, which represents the total number of hits, is divided into two parts by the dotted line. The area of the rectangular region on the right of the dotted line and the area under the curve on the left of the dotted line may be calculated as follows.

$$\begin{aligned}
\frac{l}{2}\left\{\frac{\pi}{2} - \sin^{-1}\left(\frac{l}{k}\right)\right\} + \int_0^{\sin^{-1}(l/k)} x\, d\theta &= \frac{l\pi}{4} - \frac{l}{2}\sin^{-1}\left(\frac{l}{k}\right) + \int_0^{\sin^{-1}(l/k)} \frac{k}{2}\sin\theta\, d\theta \\
&= \frac{l\pi}{4} - \frac{l}{2}\sin^{-1}\left(\frac{l}{k}\right) - \frac{k}{2}\cos\theta\Big|_0^{\sin^{-1}(l/k)} \\
&= \frac{l\pi}{4} - \frac{l}{2}\sin^{-1}\left(\frac{l}{k}\right) - \frac{k}{2}\cos\theta_1 + \frac{k}{2}
\end{aligned}$$

where $\theta_1 = \sin^{-1}(l/k)$. So,

$$\cos\theta_1 = \sqrt{1-\sin^2\theta_1} = \sqrt{1-\frac{l^2}{k^2}} = \frac{\sqrt{k^2-l^2}}{k}$$

Thus, the area under the curve is given by $\frac{l\pi}{4} - \frac{l}{2}\sin^{-1}(\frac{l}{k}) - \frac{\sqrt{k^2-l^2}}{2} + \frac{k}{2}$ and the probability of a needle hitting a line is

$$\frac{\frac{l\pi}{4} - \frac{l}{2}\sin^{-1}(\frac{l}{k}) - \frac{\sqrt{k^2-l^2}}{2} + \frac{k}{2}}{l\pi/4}$$

Investigation 1.15 The problem needs investigation only when the diameter d of the coin satisfies $d < l$. Let $x =$ distance of the centre of the coin from the nearest line. Clearly, $0 \leq x \leq l/2$. If the centre of the coin lies on one of the lines then $x = 0$ and if the centre lies exactly halfway between two lines, then $x = l/2$.

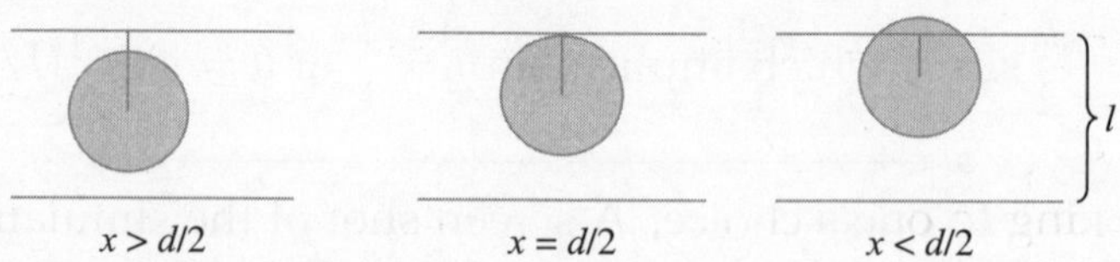

The coin will touch a line or cross over if $x \leq d/2$. These will be referred to as 'hits'. However, if $x > d/2$ then the coin will fall between two lines. Such a throw is referred to as a 'miss'.

We visualise the throw as leading to selecting a number x between 0 and $l/2$. Numbers between 0 and $d/2$ represent a hit. Hence,

$$\text{Probability of a hit} = \frac{d/2}{l/2} = \frac{d}{l}$$

B.2 Chapter 2

Investigation 2.1 The outcomes of the sample space are as follows

$$\begin{aligned} S = \{&(1,1,2),(1,1,3),(1,2,3),(1,3,2),\\ &(2,2,1),(2,2,3),(2,1,3),(2,3,1),\\ &(3,3,1),(3,3,2),(3,1,2),(3,2,1)\} \end{aligned}$$

The triples in which the first two elements are the same, represent outcomes with a probability of 1/18 each. These represent the cases where the player's choice is the same as the door which has the car behind it (and hence the player will win by sticking to his choice). The remaining outcomes represent the cases when the player's choice and location of the car are different and these have a probability of

1/9 each. In these cases the player will win if he switches. Thus, the probability of winning by sticking to the initial choice is $6 \times \frac{1}{18} = \frac{1}{3}$ and the probability of winning by switching is $6 \times \frac{1}{9} = \frac{2}{3}$.

Investigation 2.2 The steps of simulation are much the same as described earlier in the chapter.

Step 1: Enter = `RANDBETWEEN(1,3)` in cell `A2` and drag till cell `A101`. This simulates the door behind which the car is placed.

Step 2: Enter 2 in cell `B2` and drag till `B101` or simply double click on the corner of cell `B2`. This fixes the player's choice as door 2.

Step 3: To simulate the door opened by Monty we enter `=IF(A2=2, IF(RAND()<0.5, 1, 3), IF(A2=1, 3, 1))` in cell `C2` followed by a double click on the corner of the cell.

Changing the player's choice does not impact the probability of winning by switching or sticking to one's choice. A screen shot of the simulation is provided in Fig. B.2.1.

Location of car	Player's Choice	Door opened by Monty	Win by switching from original choice	Win by staying with original choice		Probability of winning by switching	Probability of winning by staying
1	2	3	YES	NO		0.67	0.33
1	2	3	YES	NO			
2	2	2	NO	YES			
2	2	3	NO	YES			
2	2	2	NO	YES			
1	2	3	YES	NO			
3	2	2	YES	NO			
1	2	3	YES	NO			
1	2	3	YES	NO			
3	2	2	YES	NO			
1	2	3	YES	NO			
2	2	3	NO	YES			
3	2	2	YES	NO			
3	2	2	YES	NO			

Fig. B.2.1 A simulation for the player's initial choice of door 2

Investigation 2.3 The steps of simulation will remain the same. However, the conditional `IF` statement in step 3 becomes `=IF(A2=1, IF(RAND()<0.25, 2, 3), IF(A2=2, 3, 2))`.

Note that the player's choice is door 1. If the car is also behind door 1, then Monty will open door 2 with a probability of 1/4 and door 3 with a probability of 3/4. To incorporate this, we generate a random number using `RAND()`. If it is less than or equal to 0.25, the output is 2 and if it is greater than 0.25, the output is 3. The rest of the statement remains the same as before. Here is a screenshot of the simulation.

Location of car	Player's Choice	Door opened by Monty	Win by switching from original choice	Win by staying with original choice		Probability of winning by switching	Probability of winning by staying
2	1	3	YES	NO		0.63	0.37
2	1	3	YES	NO			
2	1	3	YES	NO			
2	1	3	YES	NO			
3	1	2	YES	NO			
2	1	3	YES	NO			
2	1	3	YES	NO			
1	1	3	NO	YES			
1	1	3	NO	YES			
1	1	2	NO	YES			
3	1	2	YES	NO			
3	1	2	YES	NO			
3	1	2	YES	NO			
2	1	3	YES	NO			

Fig. B.2.2 A simulation showing Monty's preference for door 3 with probability 3/4

Monty's preference towards door 3 has not impacted the probability of winning by switching or staying with one's choice. The following tree diagram helps us to confirm our finding. The probabilities along the branches of the tree indicate that if the player's initial choice is door 1 then Monty will open door 2 with a probability of 1/4 and door 3 with a probability of 3/4. Similarly, the probabilities of Monty opening doors 1 and 3 are 1/4 and 3/4 respectively, when the player's initial choice is door 2. However, if the player's initial choice is door 3, then Monty will open door 1 or 2 with a probability of 1/2 each. Thus, the outcome $(1,1,2)$ can occur with a probability of $\frac{1}{3} \times \frac{1}{3} \times \frac{1}{4} = \frac{1}{36}$, whereas the probability of the occurrence of the outcome $(1,1,3)$ is $\frac{1}{3} \times \frac{1}{3} \times \frac{3}{4} = \frac{1}{12}$. However, the probabilities of the outcomes $(1,2,3)$ and $(1,3,2)$ are $\frac{1}{3} \times \frac{1}{3} \times 1 = \frac{1}{9}$ as before. It can be easily verified that the probability of winning by staying with one's original choice is $\frac{1}{3}$ and of winning by switching is $\frac{2}{3}$.

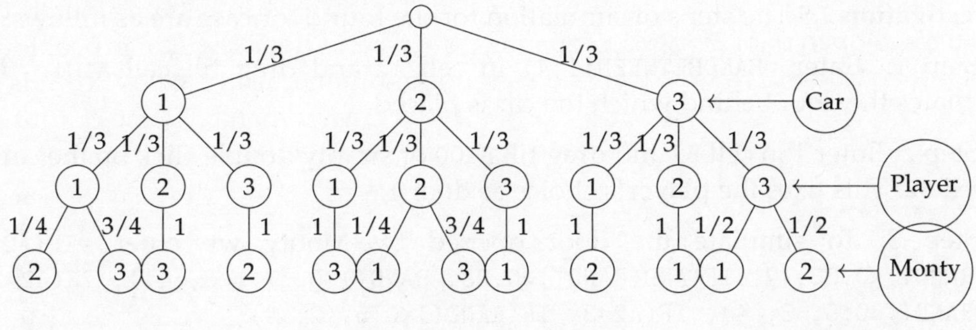

Fig. B.2.3 Tree diagram representing Monty's preference for door 3 with probability 3/4

Investigation 2.4 The following tree diagram helps us to analyse the problem for the four door case.

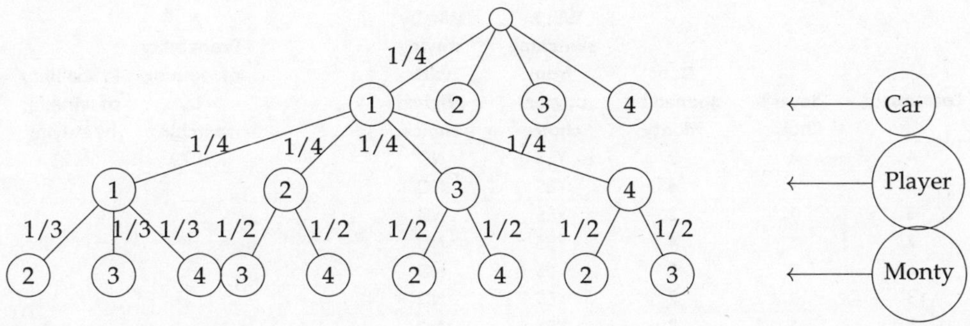

Fig. B.2.4 Tree diagram representing the 4-door case of the Monty Hall Problem, with the car behind door 1

There are 36 outcomes in the sample space. These are as follows

$$\begin{aligned} S = \{&(1,1,2),(1,1,3),(1,1,4),(1,2,3),(1,2,4),(1,3,2),(1,3,4),(1,4,2),(1,4,3),\\ &(2,2,1),(2,2,3),(2,2,4),(2,1,3),(2,1,4),(2,3,1),(2,3,4),(2,4,1),(2,4,3),\\ &(3,3,1),(3,3,2),(3,3,4),(3,1,2),(3,1,4),(3,2,1),(3,2,4),(3,4,1),(3,4,2),\\ &(4,4,1),(4,4,2),(4,4,3),(4,1,2),(4,1,3),(4,2,1),(4,2,3),(4,3,1),(4,3,2)\} \end{aligned}$$

The occurrences of each outcome of the sample space in which the first two numbers are equal have a probability of 1/48. These represent the cases when the player's choice and location of car coincide. However, the remaining cases, which represent the possibility of the player's choice being different from the location of the car, have probability 1/32. Thus, the probability of winning by sticking to the initial choice is $12 \times \frac{1}{48} = \frac{1}{4}$ and the probability of winning by switching is $24 \times \frac{1}{32} = \frac{3}{4}$.

Investigation 2.5 The steps of simulation for the four door case are as follows:

Step 1: Enter `=RANDBETWEEN(1,4)` in cell `A2` and drag till cell `A101`. This simulates the door behind which the car is placed.

Step 2: Enter 1 in cell `B2` and drag till `B100` or simply double click on the corner of cell `B2`. This fixes the player's choice as door 1.

Step 3: To simulate the door opened by Monty we enter `=IF(A2=1, IF(RAND()<0.33, 2, IF(AND(RAND()>0.33, RAND()<0.67), 3, 4)), IF(A2=2, IF(RAND()<0.5, 3, 4), IF(A2=3, IF(RAND()<0.5, 2, 4), IF(RAND()<0.5, 2, 3))))` in cell `C2` followed by a double click on the corner of the cell.

The simulation may be completed using steps 4 and 5 as in the three door case. A screenshot of the simulation is as follows.

Location of car	Player's Choice	Door opened by Monty	Win by switching from original choice	Win by staying with original choice		Probability of winning by switching	Probability of winning by staying
4	1	2	YES	NO		0.79	0.21
3	1	4	YES	NO			
3	1	2	YES	NO			
2	1	4	YES	NO			
4	1	2	YES	NO			
3	1	2	YES	NO			
3	1	2	YES	NO			
4	1	3	YES	NO			
3	1	2	YES	NO			
4	1	2	YES	NO			
4	1	3	YES	NO			
2	1	4	YES	NO			
3	1	4	YES	NO			
4	1	3	YES	NO			

Fig. B.2.5 A simulation of the four door case

Investigation 2.6 If there are 5 doors, the probability of winning by switching is 4/5 and the probability of winning by staying with one's original choice is 1/5. If the problem is generalised to n doors, the probabilities of winning by switching and staying are $(n-1)/n$ and $1/n$ respectively.

Investigation 2.7 Let us consider the possibility of obtaining a sum of 9 in the roll of a pair of dice. This can occur if the dice show up the numbers (3,6), (4,5), (5,4) or (6,3). Thus, there are four outcomes that are favourable to the occurrence of the sum of 9 and the probability of obtaining this sum is $4/36 = 1/9$. The probabilities of the other sums may be obtained in a similar manner.

Sum	2	3	4	5	6	7	8	9	10	11	12
Probability	$\frac{1}{36}$	$\frac{2}{36}$	$\frac{3}{36}$	$\frac{4}{36}$	$\frac{5}{36}$	$\frac{6}{36}$	$\frac{5}{36}$	$\frac{4}{36}$	$\frac{3}{36}$	$\frac{2}{36}$	$\frac{1}{36}$

Investigation 2.8 The following table lists the probabilities of winning a game of Craps by getting different sums in the initial roll of the pair of dice. The total probability of winning a single game of craps actually works out to approximately 0.49!

Initial total	**Probability**
4	0.027
5	0.044
6	0.063
7	0.167
8	0.063
9	0.044
10	0.027
11	0.056
Total probaility of winning	**0.491**

Fig. B.2.6 Probability of winning a single game of Craps

Investigation 2.9 If we simulate the game of Craps, we are likely to encounter the following four cases:

Case 1: The player wins as the sum of the die rolls in the first roll is 7. In this case the game doesn't continue beyond the first roll. The output from row 2 onwards may be ignored as these rows are not relevant.

	Die 1	Die 2	Sum	Win (yes = 1, No = 0)	Lose (yes = 1, No = 0)	Game continues?			
1	6	1	7	1	0	NO		Win on first roll	YES
2	6	3	9						
3	1	4	5					Lose on first roll	
4	2	6	8						
5	6	6	12					Win on point	
6	5	3	8						
7	6	2	8					Lose on point	
8	6	5	11						
9	1	4	5						
10	1	1	2						

Fig. B.2.7 A simulation showing that player wins on the first roll

Case 2: The player loses the game in the first roll, as the sum of the die rolls is 3 (the same will happen if the player gets 2 or 12). Once again the game doesn't continue beyond the first roll. As in the previous case, the output from row 2 onwards may be ignored.

	Die 1	Die 2	Sum	Win (yes = 1, No = 0)	Lose (yes = 1, No = 0)	Game continues?			
1	1	2	**3**	**0**	**1**	NO		**Win on first roll**	
2	1	2	3						
3	2	4	6					**Lose on first roll**	YES
4	5	4	9						
5	1	1	2					**Win on point**	
6	2	6	8						
7	4	4	8					**Lose on point**	
8	2	1	3						
9	5	2	7						
10	1	5	6						

Fig. B.2.8 A simulation showing that player loses on the first roll

Case 3: The sum of die rolls is 8. This means that the player neither wins nor loses in the first roll. 8 is referred to as the player's *point*. The game continues till either 8 or 7 appears. Since 8 appears in the 8th roll, the player *wins* the game in this roll. The remaining rows of the output may therefore be ignored. A similar situation may arise if the player's first roll yields a sum of 4, 5, 6, 9 or 10.

	Die 1	Die 2	Sum	Win (yes = 1, No = 0)	Lose (yes = 1, No = 0)	Game continues?			
1	6	2	**8**	**0**	**0**	YES		**Win on first roll**	
2	5	4	9	0	0	YES			
3	2	3	5	0	0	YES		**Lose on first roll**	
4	5	1	6	0	0	YES			
5	4	6	10	0	0	YES		**Win on point**	YES
6	3	6	9	0	0	YES			
7	3	1	4	0	0	YES		**Lose on point**	
8	5	3	8	1	0	NO			
9	2	4	6						
10	2	4	6						

Fig. B.2.9 A simulation showing that the player wins on a point of 8

Case 4: In this simulation, the player's point in the first roll is 8. Also 7 appears before 8 in the subsequent rolls (in fact it appears in the 15th roll) and the player *loses* the game.

	Die 1	Die 2	Sum	Win (yes = 1, No = 0)	Lose (yes = 1, No = 0)	Game continues?		
1	4	4	**8**	**0**	**0**	YES	**Win on first roll**	
2	3	3	6	0	0	YES		
3	6	4	10	0	0	YES	**Lose on first roll**	
4	5	4	9	0	0	YES		
5	3	2	5	0	0	YES	**Win on point**	
6	1	4	5	0	0	YES		
7	5	1	6	0	0	YES	**Lose on point**	YES
8	2	4	6	0	0	YES		
9	5	4	9	0	0	YES		
10	3	6	9	0	0	YES		
11	6	5	11	0	0	YES		
12	2	1	3	0	0	YES		
13	1	2	3	0	0	YES		
14	3	3	6	0	0	YES		
15	2	5	7	0	1	NO		
16	1	6	7					

Fig. B.2.10 A simulation showing that the player loses on a point of 8

B.3 CHAPTER 3

Investigation 3.1

(a) The answer is always yes, if just one extra edge is added between any two vertices.

(b) Two vertices would have same edges as before. Successful tracing will start with one of these vertices and end at the other vertex.

(c) If you change the starting vertex to any of the vertices between which an extra edge was added then no successful tracing is possible but starting at any of the unchanged vertices will give a successful tracing ending at the other.

(d) Any two of the four vertices.

(e) None.

Investigation 3.2 Answers are the same as in the previous investigation.

Investigation 3.3 Each vertex has degree 5. To find the same using `Mathematica` we first specify the graph,

```
FLR = EdgeTaggedGraph[{1<->2, 2<->3, 3<->4, 4<->1, 1<->3, 2<->4, 1<->2,
 3<->4, 1<->3, 2<->4}->{"a","b","c","d","e","f","g","h","i","j"},
 VertexLabels->"Name", EdgeLabels->"Name", EdgeStyle->{Thick}]
```

and then follow it with `VertexDegree[FLR]` which gives output $\{5,5,5,5\}$.

Investigation 3.4

(a) The original vertex degrees in KBP are $\{5,3,3,3\}$. Suppose we add an extra edge between vertex 1 and vertex 4, then the new vertex degrees will be $\{6,3,3,4\}$.

(b) The original vertex degrees in FLR are $\{5,5,5,5\}$. Suppose we add an extra edge between vertex 1 and vertex 4, then the new vertex degrees will be $\{6,5,5,6\}$.

To do the same using `Mathematica`, graphs '`KBPE`' and '`FLRE`' with the extra edge between vertices 1 and 4 will have to be specified followed by the command to find vertex degrees.

Investigation 3.5

(a) You can do it!

(b) '`KBPE`' and '`FLRE`' will be examples of such graphs.

(c) No. The sum of the degrees of the vertices will always be even. Every edge has two vertices attached to it and so it will contribute to the vertex degree of both vertices.

Investigation 3.6 For the graphs drawn in the above investigation as responses to Part (b), the answer will be yes. For the graphs in Part (a) the answer is no.

Investigation 3.7 Try

```
conng = EdgeTaggedGraph[{1<->2, 2<->3, 3<->1, 3<->4},
 VertexLabels ->{1->u, 2->v, 3->w, 4->z},
 EdgeStyle ->{Thick}]
```

FLR is not a simple graph as there are pairs of vertices with multiple edges. The graph in Fig. 3.2.9 is simple. The vertex and edge sets are given by $V = \{u,v,w,z\}$ and $E = \{uv, vw, wu, wz\}$.

Investigation 3.8 Let $V = \{1,2,3\}$. Then the different possible simple graphs with vertex set V will be the graphs with vertex set V and edge set E where $E \subseteq \{12, 13, 23\}$. Since there will be 8 possibilities for E, we will correspondingly get 8 simple graphs on three vertices.

One way of systematically drawing these simple graphs would be using the `EdgeAdd` command as shown below. We first define a graph '`g0`' with three vertices and no edges:

```
g0 = Graph[{1,2,3}, {}, VertexLabels->"Name", EdgeStyle->{Thick}]
```

If we follow it with `g12 = EdgeAdd[g0, 1<->2]` we get a graph 'g12' with vertex set $V = \{1,2,3\}$ and edge set $E = \{12\}$. We similarly create graphs 'g23' and 'g13' which have edge sets $\{23\}$ and $\{13\}$ respectively.

Using `EdgeAdd` on 'g12', we can get two more graphs with edge sets $\{12,23\}$ and $\{12,13\}$ respectively. Adding the edge `1<->3` to 'g23', gives us the third simple graph with edge set of size 2. The eighth and last simple graph can be obtained by adding one more edge to any of the graphs found at the previous stage.

Investigation 3.9 The following code produces the graph in Fig. 3.3.2.

```
graph1 = EdgeTaggedGraph[{1<->2, 1<->3, 2<->3, 3<->1, 3<->4, 4<->4},
 VertexLabels->{1->u, 2->v, 3->w, 4->z}, EdgeStyle->{Thick}]
```

To convert the above graph into a simple graph we need to remove the loop on vertex z and also one of the two edges between vertices u and w:

```
conng = EdgeTaggedGraph[{1<->2, 1<->3, 2<->3, 3<->4},
 VertexLabels->{1->u, 2->v, 3->w, 4->z}, EdgeStyle ->{Thick}]
```

Note that this is the same as the graph in Fig. 3.2.9.

Investigation 3.10 Let $V = \{u,v\}$. We want to draw graphs G with Vertex set V and edge set E where $|E| = 3$. We are allowed loops and multiple edges. A systematic way of doing this would be to keep count of the number of loops at the vertices u and v. Using this let us consider the set `EdgeInfo` $= \{(x,e,y) \mid x+e+y = 3; x,e,y \geq 0\}$ where each triple represents the following data for a given graph—x is the number of loops at vertex u, the number e is the number of edges between u and v, and y is the number of loops at vertex v. Since the number of edges on these graphs has to be 3, then $x+e+y = 3$ and each of x,y and e is a non-negative integer.

Clearly the graphs we draw will be in one-to-one correspondence with the elements of `EdgeInfo`. It is not difficult to see that the elements of `EdgeInfo` are

$$\{(3,0,0),(2,1,0),(2,0,1),(1,2,0),(1,0,2),(1,1,1),(0,3,0),(0,0,3),(0,1,2),(0,2,1)\}.$$

Some of the corresponding graphs are displayed in Fig. B.3.1.

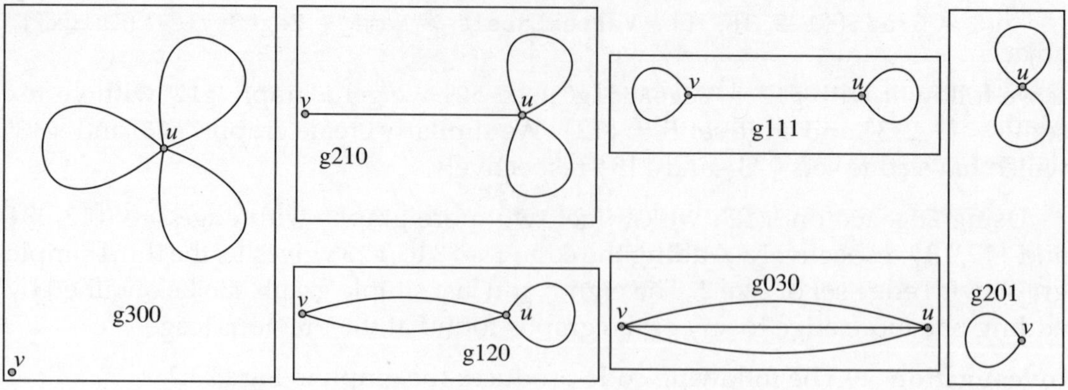

Fig. B.3.1 Some graphs on 2 vertices with 3 edges

If we have no condition on the size of the edge set, then we can draw infinitely many graphs on two vertices.

Investigation 3.11 All the graphs referred to are connected. The KBP graph can be modified to a disconnected graph by removing the 3 edges incident at vertex C. Also if vertices A and D are removed from the KBP graph then we are left with a graph with two isolated vertices which is clearly disconnected.

The FLR can be modified to a disconnected graph by removing all 5 edges incident to any chosen vertex. The only way to modify FLR into a disconnected graph will be by removing 3 of the vertices.

The single vertex simple graph in Fig. 3.3.1 will always be connected. The second graph in the figure can be modified to a disconnected graph only be removing the edge between vertices b and c. The third graph in the collection can be modified to give a disconnected graph by either removing one of the two edges, or by removing the vertex v.

The graph in Fig. 3.3.2 can be modified to a disconnected graph by either removing vertex w or by removing the edge between vertices w and z.

Investigation 3.12 The graph in Fig. 3.3.2 can be produced as follows:

```
graph1 = EdgeTaggedGraph[{1<->2, 1<->3, 2<->3, 3<->1, 3<->4, 4<->4},
 VertexLabels->{1->u, 2->v, 3->w, 4->z}, EdgeStyle->{Thick}]
```

The command `VertexDegree[graph1]` gives the output $\{3,2,4,3\}$ which tells us that vertex u has degree 3, vertex v has degree 2, vertex w has degree 4 and vertex z has degree 3.

Now `AdjacencyList[graph1, 1]` gives the output $\{2,3\}$ which says that u is adjacent to v and w. Similarly we can see using `Mathematica` or directly that v is adjacent to u and w. The vertex w is adjacent to u, v and z and the vertex z is adjacent to w. (Even though there is a loop incident on z, usually adjacency is defined between two distinct vertices.)

Adjacent edges can be found using `IncidenceList[graph1, v]`. This gives the output of all edges that are incident at vertex v. Hence any pair of such edges will be considered adjacent. For example, `IncidenceList[graph1, 1]` gives

$$\left\{1 \xrightarrow{1} 2,\ 1 \xrightarrow{1} 3,\ 3 \xrightarrow{1} 1\right\}$$

Fig. B.3.2 Incidence list of Graph I at Vertex u

Let us denote the edge uv by a and the two edges joining u and w as b and c respectively. Then, the above output is telling us that any two of a, b and c are adjacent edges. In this way, by finding incidence lists for each vertex we can find all the adjacent edges. We will also get that the edge wz and the loop zz are adjacent.

There is no end-vertex in the graph in Fig. 3.3.2 as there is no vertex of degree 1. The sum of the degrees of all the vertices is $3 + 2 + 4 + 3 = 12$. The edge set has size 6. We see that the sum of degrees of all vertices is twice the number of edges. (See the discussion for Investigation 3.5.)

Investigation 3.14 See the text following the statement of this Investigation.

Investigation 3.15 See the text following the statement of this Investigation.

Investigation 3.16 The number of arrangements for three discs equals $3 \times 3 \times 3 = 3^3$. Generalising this, the number of arrangements for n discs is 3^n. For d discs on p pegs the total number of arrangements is p^d. The first disc can be placed in p ways, the second disc can also be placed in p ways, and so on, till the last disc.

Investigation 3.17 01 and 02 can be reached from 00 in one move. We can also obtain 02 from 01 in one move and vice versa. Similarly 10 and 12 can be reached from 11 in one move. 10 can be reached from 12 in one move and vice versa. 20 and 21 can be reached from 22 in one move. Further 20 can be reached from 21 in one move and vice versa.

Investigation 3.18 The Hanoi graph $H(3,3)$ will have 27 vertices in all. The homogeneous vertices 000, 111 and 222 will be placed in the top, left and right corners. Further the three copies of $H(2,3)$ may be positioned in the top, left and right corners. As in $H(2,3)$, the three homogeneous vertices lead to only two other states whereas the heterogeneous vertices lead to three other states.

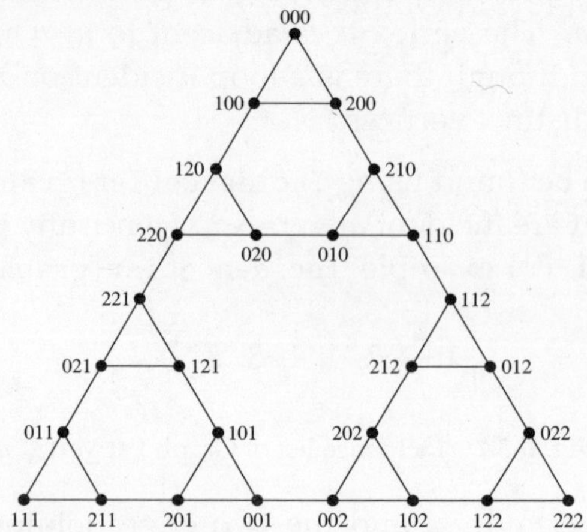

Fig. B.3.3 H(3,3) created using three copies of H(2,3)

Investigation 3.19 A path between the homogeneous vertices 000 and 111 in $H(3,3)$ is 000 $\rightarrow$ 001 $\rightarrow$ 021 $\rightarrow$ 022 $\rightarrow$ 122 $\rightarrow$ 120 $\rightarrow$ 110 $\rightarrow$ 111. It comprises 7 edges which correspond to the minimum number of moves for shifting the three discs from peg 0 to peg 1. A similar path comprising 15 edges between the homogeneous vertices 0000 and 2222 in $H(4,3)$ is 0000 $\rightarrow$ 0001 $\rightarrow$ 0021 $\rightarrow$ 0022 $\rightarrow$ 0122 $\rightarrow$ 0120 $\rightarrow$ 0110 $\rightarrow$ 0111 $\rightarrow$ 2111 $\rightarrow$ 2112 $\rightarrow$ 2102 $\rightarrow$ 2100 $\rightarrow$ 2200 $\rightarrow$ 2201 $\rightarrow$ 2221 $\rightarrow$ 2222.

B.4 Chapter 4

Investigation 4.1 Uttara has sent Neelambari a locked box to which Neelambari does not have a key. Neelambari is not allowed to break the box. She adds her own lock in addition to that put by Uttara and then sends the box, now locked with two locks, back to Uttara. When she receives the box, Uttara unlocks her lock with her key. The box is still locked but with Neelambari's lock. Uttara posts the locked box to Neelambari. When she receives it, Neelambari unlocks the box with the key to her lock.

To send a secret message, you have to encrypt it. This is like locking the box with the message in it using the encryption key. However to unlock the box, that is to decrypt the message, the decryption key is required which is also secret and so has to be sent again via encryption or equivalently in a locked box. The lock-exchange done by Uttara and Neelambari is the physical equivalent of the Diffie–Helman Key exchange protocol.

Investigation 4.2

(a) (i) Input: `Times[43,53]` Output: 2279

(ii) Input: `FactorInteger[3869]`

Output: $\{\{53,1\},\{73,1\}\}$

So $3869 = 53 \times 73$.

(b) Input: `FactorInteger[235764327]`

Output: $\{\{3,1\},\{431,1\},\{182339,1\}\}$

So $235764327 = 3 \times 431 \times 182339$.

Investigation 4.3

Input: `Mod[3^7 + (-5)^51, 7]`

Output: 4

Investigation 4.4

(a) Input: `Mod[1051 + 2733, 13]`

Output: 1

Input: `Mod[Mod[1051, 13] + Mod[2733, 13], 13]`

Output: 1

(b) Input: `Mod[1051*2733, 13]`

Output: 1

Input: `Mod[Mod[1051, 13]*Mod[2733, 13], 13]`

Output: 1

(c) Input: `PowerMod[1051, 5, 13]`

Output: 7

Input: `PowerMod[Mod[1051, 13],5, 13]`

Output: 7

Investigation 4.5

Input: `EulerPhi[Times[53, 37, 24]]`

Output: 14976

Investigation 4.6

(a) Input: `FactorInteger[596232]`

Output: $\{\{2,3\},\{3,2\},\{7,2\},\{13,2\}\}$

So $596232 = 2^3 \times 3^2 \times 7^2 \times (13)^2$

(b) Input: `EulerPhi[596232]`

Output: 157248

(c) Input: `Times[2^3 -2^2, 3^2-3, 7^2-7, 13^2 - 13]`

Output: 157248

Investigation 4.7

Manual method for finding the inverse of 75 modulo 7.

Step 1: $75 = (10 \times 7) + 5$

Step 2: $7 = (1 \times 5) + 2$

Step 3: $5 = (2 \times 2) + 1$

Step 4: $1 = 5 - (2 \times 2)$ by rewriting Step 3

Step 5: $1 = 5 - (2 \times (7 - 5)) = (-2 \times 7) + (3 \times 5)$, using Step 2 to substitute for 2

Step 6: So $1 = (-2 \times 7) + (3 \times (75 - (10 \times 7))$, using Step 1 to substitute for 5

Step 7: $1 = (3 \times 75) + (-32 \times 7)$

Step 8: Thus $75 \times 3 = 1 \bmod 7$ or the inverse of 75 modulo 7 is 3

Finding the inverse of 75 modulo 7 using `QuotientRemainder[a,n]`. Steps 1–3 of above can be done using this code. Steps 4–8 are as above.

- Input: `QuotientRemainder[75, 7]` gives Output: $\{10,5\}$. So $75 = (10 \times 7) + 5$.
- Input: `QuotientRemainder[7, 5]` gives Output: $\{1,2\}$. So $7 = (1 \times 5) + 2$.
- Input: `QuotientRemainder[5, 2]` gives Output: $\{2,1\}$. So $5 = (2 \times 2) + 1$.

Also the Input: `ModularInverse[75,7]` returns the Output: 3.

Investigation 4.8

(a) Input: `FactorInteger[9091]`

Output: $\{\{9091,1\}\}$

Thus the prime factorisation of 9091 is given as 9091. So 9091 is prime.

(b) We have to check if $1 = (11)^{9091} \bmod 9091$. So we

Input: `PowerMod[11,9090, 9091]` to get Output: 1 as required.

(c) Input: `FactorInteger[100001]`

Output: $\{\{11, 1\}, \{9091, 1\}\}$

So $100001 = 11 \times 9091$.

(d) To find $\phi(100001)$ we input: `EulerPhi[100001]` to get output: 90900. Thus $\phi(100001) = 90900$.

Now using *Mathematica*, `EulerPhi[11]`$=$ 10 and `EulerPhi[9091]`$=$ 9090. Thus `EulerPhi[100001]` $= 90900 = 10 \times 9090 =$`EulerPhi[11]`$\times$`EulerPhi[9091]`. Thus $\phi(11 \times 9091) = \phi(11) \times \phi(9091)$.

(e) Since $n = 100001 = 11 \times 9091$ is the prime factorisation for n, we have $\gcd(53, 100001) = 1$. Thus $a = 53$ is coprime to n. We need to verify that $1 = a^{\phi(n)} \bmod n$.

So we input: `PowerMod[53, 90900, 100001]` to get output: 1.

Investigation 4.9

(a) Input: `GCD[560027, 1874]` gives Output: 1.

(b) Input: `ModularInverse[1874,560027]` returns Output: 132984.

(c) Input: `FactorInteger[560027]` returns Output: $\{\{13, 1\}, \{23, 1\}, \{1873, 1\}\}$. So $560027 = 13 \times 23 \times 1873$ as its prime factorisation.

(d) Input: `EulerPhi[560027]` gives Output: 494208. So $\phi(560027) = 494208$. Further `Times[12, 22, 1872]`$=$ 494208 or we have $\phi(13) \times \phi(23) \times \phi(1873) = 12 \times 22 \times 1872 = 494208 = \phi(560027)$

(e) Since $a = 1874$ is coprime to $n = 560027$, by Euler's Theorem we should have $1 = a^{\phi(n)} \bmod n$. Now if we Input: `PowerMod[1874, 494208, 560027]` we get the Output: 1. So Euler's Theorem is verified.

Investigation 4.10 Let $n = 2279$ be a product of two distinct primes p, q. Let $\mathbf{e} = 73$. Then

(a) `EulerPhi[2279]`$=$ 2184. So $\phi(2279) = 2184$.

(b) We know that if $n = 2279 = pq$ then $\phi(n) = (p-1)(q-1) = pq - (p+q) + 1$. Thus we get

$$\begin{aligned} p + q &= 2279 - 2184 + 1 = 96 \\ pq &= 2279 \end{aligned}$$

So p, q are roots of the quadratic equation $x^2 - 96x + 2279 = 0$. So we can take $p = \frac{1}{2}(96 + \sqrt{(96^2 - 4 \times 2279)}$ and $q = \frac{1}{2}(96 - \sqrt{(96^2 - 4 \times 2279)}$.

Using `Mathematica` we get `Sqrt[96^2-4*2279]`$= 10$. Thus $p = 106/2 = 53$ and $q = 86/2 = 43$.

(c) Input of `FactorInteger[2279]` gives Output: $\{\{43, 1\}, \{53, 1\}\}$. Hence we see that p, q are as above.

(d) Now $\mathbf{d}$ is the inverse of $\mathbf{e} = 73$ modulo $\phi(2279) = 2184$. Thus we have `ModularInverse[73, 2184]`$= 1825 = \mathbf{d}$.

Investigation 4.11 Note that the RSA Public Directory has the following listed for YOU: the public $n = 123889$ and $\mathbf{e} = 71$. Cheeni has sent $C = 10103911264451987$.

Our first task is to find the corresponding ciphertext. For this, we have to make blocks from the left choosing as many digits to get the largest number less than 257. Thus the ciphertext number blocks for ASCII are 101 039 112 64 45 198 7. We use the ASCII Table to get the ciphertext e'p@-ã. Note that 07 is control sequence BEL and so is not represented in the ciphertext.

Now we recover the plaintext number M and the plaintext (secret message) from C using the RSA process. The first step involves making blocks for C, starting left to right, choosing as many digits to get the largest number less than $n = 123889$.

The blocks we get are $C_1 = 101039, C_2 = 112644, C_3 = 51987$. Since

$$M_i = C_i^{\mathbf{d}} \bmod n = \texttt{PowerMod[}C_i\texttt{, 102311, 123889]}$$

we get $M_1 = 77696, M_2 = 98432$, and $M_3 = 7769$.

So $M = 77696984327769$. We convert this to plaintext. The blocks for conversion using the ASCII Table are 77 69 69 84 32 77 69 and so we recover the message or plaintext as MEET ME.

Now we have to verify that if we start with plaintext MEET ME and complete the RSA encryption process we indeed get the C we started with.

Now the plaintext is MEET ME. So the corresponding plaintext number M obtained using the ASCII Table is $M = 77696984327769$. It is given that $n = 123889$. We make blocks by using as many digits from left to right till we get the largest number below n. So here we will get 3 blocks $M_1 = 77696$, $M_2 = 98432$ and $M_3 = 7769$. Further we know that $C_i = M_i^{\mathbf{e}} \bmod n =$ `PowerMod[`M_i`, 71, 123889]`. Thus we get $C_1 = 101039, C_2 = 112644$ and $C_3 = 51987$. Therefore $C = 10103911264451987$.

Investigation 4.12 Recall that Bulku has $n = 100761443$ and $\mathbf{e} = 1007$ in the RSA Public Directory. You want to send Bulku a secret message. The secret message is 'Run away'. Do the following.

(a) The plaintext given is RUN away. Using the ASCII Table the corresponding plaintext number is $M = 828578329711997121$.

(b) To encrypt M we need to break it into blocks, starting from left to right using as many digits to get the largest number less than $n = 100761443$. Thus we get 3 blocks namely $M_1 = 82857832, M_2 = 97119971$ and $M_3 = 21$. Now

$$C_i = M_i^{\mathbf{e}} \bmod n = \texttt{PowerMod[}M_i\texttt{, 1007, 100761443]}$$

Thus we get $C_1 = 49744979, C_2 = 63332321$ and $C_3 = 48181572$. Hence the ciphertext number $C = 497449796333232148181572$.

(c) To the ciphertext number C to ciphertext using the ASCII Table we break C into blocks. We start from left to right using as many digits to get the largest number less than 257. So the blocks for conversion to ciphertext using the ASCII Table are given by 49 74 49 79 63 33 232 148 181 57 2. The corresponding ciphertext is 1J1O?!ÞöÁ9. Note that 02 is the control character STX and so is not represented.

(d) We know that **d** is the inverse of $\mathbf{e} = 1007$ modulo $\phi(100761443)$. So we get

$$\mathbf{d} = \texttt{ModularInverse[1007, EulerPhi[100761443]]} = 54722471$$

To decrypt $C = 497449796333232148181572$ using the RSA decryption process, we first need to break it into blocks. We start from left to right choosing as many digits to get the largest number less than $n = 100761443$. Thus we get 3 blocks, namely, $C_1 = 49744979, C_2 = 63332321$ and $C_3 = 48181572$. Now

$$M_i = C_i^{\mathbf{d}} \bmod n = \texttt{PowerMod[}C_i\texttt{, 54722471, 100761443]}$$

So we get $M_1 = 82857832, M_2 = 97119971$ and $M_3 = 21$. So we get back $M = 828578329711997121$. To recover the plaintext we make blocks by using digits from let to right till we get the largest number less than 257. The corresponding blocks for using with the ASCII Table are 82 85 78 32 97 119 97 121. From the ASCII Table we convert to get RUN away.

B.5 Chapter 5

Investigation 5.1 The number of black triangles at stages 4 and 5 will be $3^4 = 81$ and $3^5 = 243$ respectively. The number of black triangles at stage n is 3^n. The shaded area at the nth stage is $(3/4)^n$.

Investigation 5.2 Stage 3 of the Sierpinski triangle will have three copies of stage 2, nine copies of stage 1 and twenty seven copies of stage 0. In general we can identify three copies of stage $n-1$ in stage n, nine copies of stage $n-2$ in stage n and so on.

Investigation 5.3 The recursive rule for the shaded area at the nth stage is $A_n = (3/4)A_{n-1}$. The explicit rule for the shaded area is $A_n = (3/4)^n$.

Investigation 5.4 As n approaches infinity, the number of shaded triangles approaches infinity while the shaded area approaches 0.

Investigation 5.5 We observe that as n increases, the number of shaded triangles increases exponentially whereas the shaded area approaches 0.

A	B	C
stage	number of black traingles	shaded area
0	1	1
1	3	0.75
2	9	0.5625
3	27	0.421875
4	81	0.31640625
5	243	0.237304688
6	729	0.177978516
7	2187	0.133483887
8	6561	0.100112915
9	19683	0.075084686
10	59049	0.056313515
11	177147	0.042235136
12	531441	0.031676352
13	1594323	0.023757264
14	4782969	0.017817948
15	14348907	0.013363461
16	43046721	0.010022596
17	129140163	0.007516947
18	387420489	0.00563771
19	1162261467	0.004228283
20	3486784401	0.003171212

Fig. B.5.1 A spreadsheet exploration of the Sierpinski Triangle

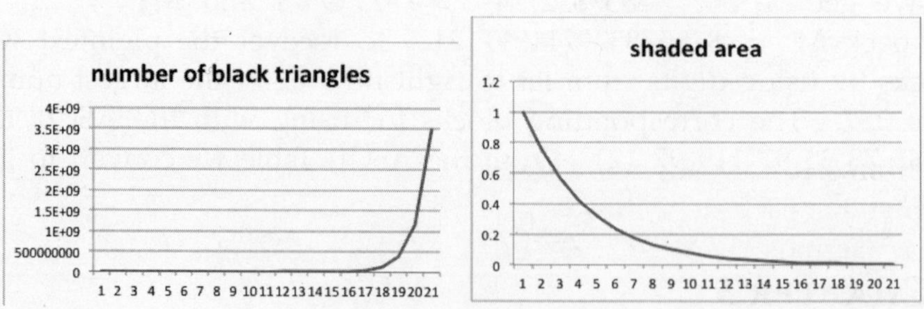

Fig. B.5.2 The number of shaded triangles approaches infinity as n increases, while the shaded area approaches 0

Investigation 5.6 Let S_n denote the number of shaded squares at stage n. Then the recursive rule for the number of shaded squares is $S_n = 8S_{n-1}$. The explicit rule is $S_n = 8^n$. The recursive rule for the shaded area is $A_n = (8/9)A_{n-1}$. The explicit formula is $A_n = (8/9)^n$.

Investigation 5.7 The number of shaded squares approaches infinity whereas the shaded area approaches 0.

Investigation 5.8 The Excel Sheet to explore the Sierpinski Carpet is as follows. In column A we represent the stage n. Column B shows the number of shaded squares. To generate this, enter the formula = `8*B1` in cell `B2`. Column C shows the shaded area at the different stages. In cell `C2` we enter the formula = `(8/9)*C1`.

A	B	C
stage	number of squares	shaded area
0	1	1
1	8	0.888888889
2	64	0.790123457
3	512	0.702331962
4	4096	0.624295077
5	32768	0.554928957
6	262144	0.493270184
7	2097152	0.438462386
8	16777216	0.389744343
9	134217728	0.346439416
10	1073741824	0.307946148

Fig. B.5.3 A spreadsheet exploration of the Sierpinski Carpet

Investigation 5.9 Each successive term of the sequence of the number of line segments may be obtained by multiplying the previous term by 4. The nth term is 3×4^n.

Stage	0	1	2	3	$\cdots$	n
No. of line segments	3	3×4	3×4^2	3×4^3	$\cdots$	3×4^n

Investigation 5.10 Each successive term of the sequence representing the length of a line segment at each stage may be obtained by multiplying the previous term by 1/3. The nth term is $(1/3)^n$.

Stage	0	1	2	3	$\cdots$	n
Length of line segment	1	1/3	$1/3^2$	$1/3^3$	$\cdots$	$1/3^n$

Investigation 5.11 The perimeter of the snowflake at each successive stage is as follows:

Stage	0	1	2	3	$\cdots$	n
Perimeter	3	$3 \times (4/3)$	$3 \times (4/3)^2$	$3 \times (4/3)^3$	$\cdots$	$3 \times (4/3)^n$

The recursive formula for the perimeter is $P_n = (4/3) \times P_{n-1}$. The explicit formula is $P_n = 3 \times (4/3)^n$.

Investigation 5.12 At stage 2, 12 small equilateral triangles each of side 1/9 get added. Hence the total area at stage 2 is $\frac{\sqrt{3}}{4} + 3 \times \frac{\sqrt{3}}{4} \times (\frac{1}{3})^2 + 12 \times \frac{\sqrt{3}}{4} \times (\frac{1}{9})^2 = \frac{10\sqrt{3}}{27}$. At stage 3, 48 smaller equilateral triangles of side 1/27 get added. Here the total area is

$$\frac{10\sqrt{3}}{27} + 48 \times \frac{\sqrt{3}}{4} \times (\frac{1}{27})^2 = \frac{282\sqrt{3}}{729}$$

Investigation 5.13 In iteration 1, the space filling tree has three branches of length 1/3 units each. In iteration 2, 12 branches of length 1/6 unit each get added. Iteration 3 has 42 new branches of length 1/12 unit each.

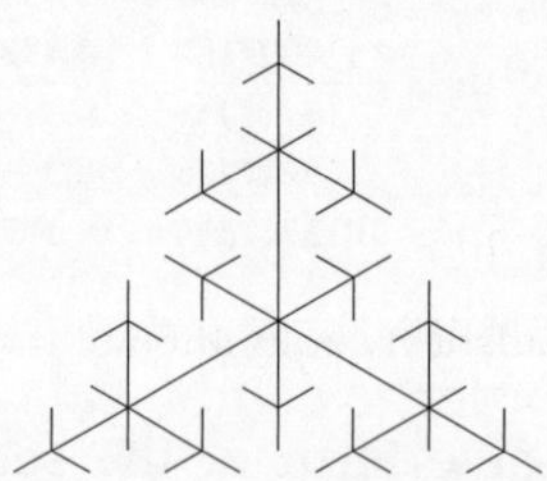

Fig. B.5.4 T_3 or stage 3 of the space filling tree

Note that each of the three side triangles contributes 12 branches each. However one branch is superimposed on an earlier branch and is not visible. Thus, the three side triangles yield a total of 11 times 3 = 33 branches. The centre triangle yields 9 branches. Hence the total number of branches added in iteration 3 is 42. Since each is of length 1/12 units the total length is $42/12 = 7/2$.

The total length of the branches in T_3 is $1 + 2 + 7/2 = 13/2$ units.

Investigation 5.14 To solve the recurrence relation, $T_{n+1} = 3T_n + 2(4^n - T_n - 4^{n-2})$ for $n > 2$ we simplify the expression to obtain $T_{n+1} = T_n + 2(4^n - 4^{n-2}) = T^n + 2 \cdot 4^{n-2}(4^2 - 1) = T_n + 30 \cdot 4^{n-2}$ for $n > 2$. Substituting $n = 3, 4, 5, \ldots$ and generalising for the nth stage, we get

$$
\begin{aligned}
T_3 &= T_2 + 30 \cdot 4^0 = T_2 + 30 \\
T_4 &= T_3 + 30 \cdot 4^1 = T_2 + 30 + 30 \cdot 4^1 \\
T_5 &= T_4 + 30 \cdot 4^2 = T_2 + 30 + 30 \cdot 4^1 + 30 \cdot 4^2 \\
&\vdots \\
T_n &= T_2 + 30 + 30 \cdot 4^1 + 30 \cdot 4^2 + \cdots + 30 \cdot 4^{n-3} \\
&= T_2 + 30(1 + 4 + 4^2 + \cdots + 4^{n-3}) \\
&= T_2 + 30 \cdot \frac{4^{n-2} - 1}{4 - 1} \\
&= T_2 + 10 \cdot (4^{n-2} - 1) \\
&= 12 + 10 \cdot (4^{n-2} - 1) \qquad \text{(since } T_2 = 12\text{)} \\
&= 2 + 10 \cdot 4^{n-2} \qquad \text{(where } n > 2\text{)}.
\end{aligned}
$$

Investigation 5.15 Let us consider a cube whose edge is of unit length. We can cut it into 8 smaller cubes, each of edge 1/2 unit. Thus $a = 8$ and $s = 1/2$. The formula $a = 1/s^D$ yields $8 = 1/(1/2)^D$ which gives $D = 3$. We may also cut the unit cube into 27 smaller cubes of edge 1/3 unit each. Substituting $a = 27$ and $s = 1/3$ gives $27 = 1/(1/3)^D$ which also gives $D = 3$!

Investigation 5.16 In stage 1 of the Sierpinski carpet, we have 8 copies of stage 0 and the side length of each smaller square is 1/3 unit (assuming that the side of the original square at stage 0 is 1 unit). Thus $a = 8$ and $s = 1/3$. Substituting in $a = 1/s^D$ we get $8 = 3^D$. In order to compute D we apply log on both sides of the equation: $\log 8 = D \log 3$, which implies $D = \frac{\log 8}{\log 3} = \frac{3 \log 2}{\log 3} \approx 1.892$. In stage 2 we have 64 smaller squares each of side length 1/9. Substituting $a = 64$ and $s = 1/9$ we get $64 = 1/(1/9)^D$ or $64 = 9^D$. Applying log on both sides like before, we obtain $D = \frac{\log 64}{\log 9} = \frac{3 \log 2}{\log 3} \approx 1.892$. Thus the self-similarity dimension of the Sierpinski carpet is approximately 1.892.

B.6 Chapter 6

Investigation 6.2 Use the given numbers for `x[1]` in the preceding code, and vary `iterations` to see more terms of the sequence. You should find the following convergences: $0 \to 0.347296$, $2 \to 1.53209$, $-2 \to -1.87939$, $1000 \to 1.53209$, $-1000 \to -1.87939$. The orbits of ± 1 are not defined since $p'(\pm 1) = 0$.

Investigation 6.3 The orbits for $x_0 = 0, 1$ cycle through 0 and 1, and do not converge to a solution. The orbit for $x_0 = 10$ eventually starts cycling through 0 and 1. The orbit for $x_0 = 5$ converges to -1.76929.

Investigation 6.4 When x_0 is just under 1.39175 the orbit converges to 0. When it is just over, the orbit diverges.

Investigation 6.5

x_0	0	0.2	0.4	0.6	0.8	1	1.2	1.4
Limit	0	0	0	1	1	1	1	1

Since the function is odd, the situation is symmetric about origin and fresh calculations are not needed for negative x.

Investigation 6.6 Use `Solve[nrq[nrq[t]]==t,t,Reals]`.

Investigation 6.7 First show that $n(x) = x\left(\dfrac{2x^2}{3x^2-1}\right)$ and then that $|x| < 1/\sqrt{5}$ implies $\left|\dfrac{2x^2}{3x^2-1}\right| < 1$.

Investigation 6.8 The graph suggests there should be a point to the left of $1/\sqrt{5}$ which is taken to -1 by Newton–Raphson. You should show this number is $1/2$. Then the orbits starting in $(1/2, 1/\sqrt{5})$ must converge to -1. Can you say anything about what happens to the left of $1/2$?

Investigation 6.13 Here, we vary the coefficient of z^2 over the values -2.5, -2.525, -2.55, -2.575 and -2.6. We see that zones of non-convergence appear, change shape and then dissipate. You should try other variations in the coefficients. And when you read the later section on quadratic iteration, compare these shapes with ones you see there.

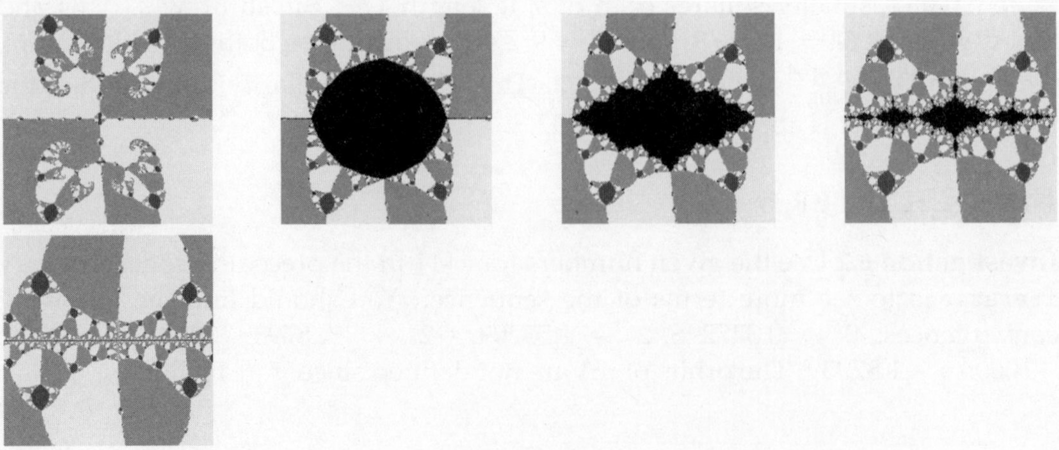

Investigation 6.14 Rearrange to $2z^3 = (3z^2 - 1)\bar{z}$. Put $z = x + iy$ and solve for x, y.

Investigation 6.15 The following code will identify 3-cycles as well as fixed points, and present them in a list called `cycles`.

```
p[z_]:=z^3-z;
nr[z_]:=z-p[z]/p'[z];
cycles=z/.NSolve[nr[nr[nr[z]]]==z, z]
```

The next piece of code takes the first member of `cycles` and shows the first 50 members of its orbit. We see the 3-cycle starts to degenerate after about 45 terms, and then converges to 1.

```
orbit[1] = cycles[[1]];
iterations = 60;
For[k=1, k<iterations, k++, orbit[k+1] = nr[orbit[k]]];
ComplexListPlot[Table[orbit[k], {k,1,iterations}], Joined->True,
 AspectRatio->0.5]
```

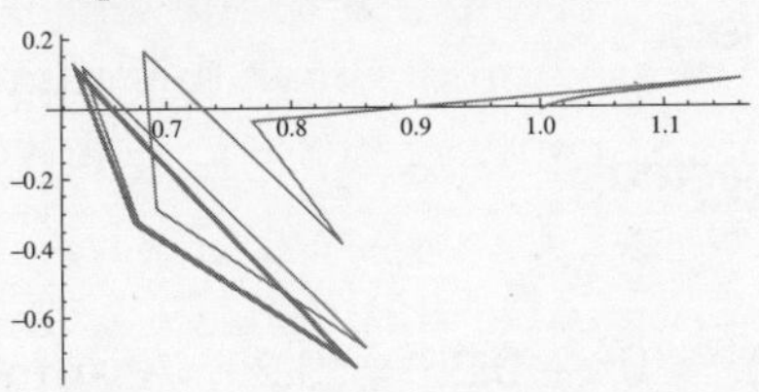

Investigation 6.18 We calculate the orbit and plot a histogram with a bin width of 0.3:

```
y[1] = 10.;
iterations = 10000;
For[k=1, k<iterations, k++, y[k+1]=(y[k]-1/y[k])/2];
Histogram[Table[y[k], {k,1,iterations}], {0.3}]
```

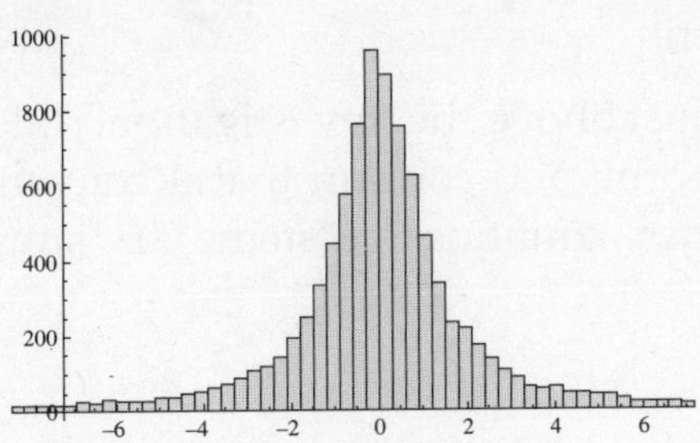

If you have studied the normal distribution in Probability, you would notice the similarity with its probability density function.

Investigation 6.19 We see a 2-cycle that degenerates into chaotic behaviour.

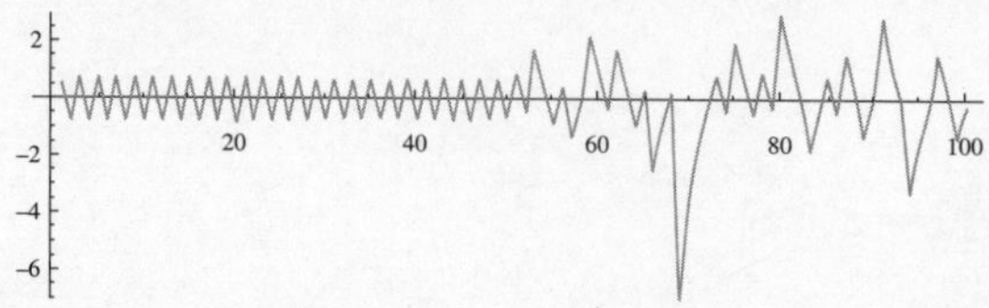

You can plot more terms and also draw a histogram to see the distribution.

Investigation 6.22 Look at c near $1/4$.

Investigation 6.23 Look at c near -2.

Investigation 6.24 Some instances:

c	Nature of orbit of 0	Connectivity of Julia set
0	Stationary at 0	Connected
0.2	Converges	Connected
0.3	Goes to infinity	Totally disconnected
$0.6i$	Converges	Connected
$0.7i$	Goes to infinity	Totally disconnected

Do you see a pattern? Try more cases!

Investigation 6.25 For $n = 1$ we have $\Phi^1_{0.25}(0) = 0^2 + 0.25 = 0.25$. Assume the induction hypothesis for n, and consider $n + 1$. Then,

$$0 < \Phi^n_{0.25}(0) < 0.5 \implies 0^2 + 0.25 < (\Phi^n_{0.25}(0))^2 + 0.25 < 0.5^2 + 0.25$$
$$\implies 0.25 < \Phi^{n+1}_{0.25}(0) < 0.5$$

Investigation 6.26 Use mathematical induction.

Investigation 6.27 Use mathematical induction.

Investigation 6.29 Apply 'completion of squares'.

Investigation 6.30 Use mathematical induction.

Investigation 6.31 Consider the bulb shown just above this investigation. Take a point in it, such as $-0.1 + 0.7i$. Compute its orbit and confirm that it converges to a 3-cycle. And the 'antenna' has 3 branches, counting the stem. Try similar calculations for other bulbs.

B.7 Chapter 7

Investigation 7.1

(a) A regular pentagon has 5 rotational symmetries. These are rotations by angles of degree 0, 72, 144, 216 and 288 about the centre of the regular pentagon in the counter-clockwise direction. So $R_{72} = (A\ B\ C\ D\ E)$ whereas $R_{216} = (A\ C\ E\ B\ D)$. Others can be written down similarly.

(b) There are 5 lines of reflection that will give us symmetries of a regular pentagon. These are the lines joining the vertex to the mid-points of the opposite sides. So if F_A is the reflection symmetry about the line joining vertex A to the mid-point of the opposite side CD, then $F_A = (A)(B\ E)(C\ D)$. Similarly the four other reflection symmetries can be written down.

(c) A regular n-gon will have $2n$ symmetries consisting of n rotations and n reflections. Can you find the angles of rotation? (Hint for a regular 5-gon it was $360k/5$ where $k = 0, 1, \ldots, 4$.) Depending on whether n is even or odd, lines of reflection symmetry will vary. For example the 4 lines of reflection that give symmetries of a square are the diagonals and lines joining mid-points of opposite sides. A regular n-gon is said to have symmetry type D_n as its group of symmetries is called the dihedral group D_n of degree n. Note that it has $2n$ elements consisting of n rotations and n reflections.

Investigation 7.2

(a) A partial Cayley table for the symmetries of an equilateral triangle is shown below. We show one method for calculating $R_{120} \circ F_A$. You can also work this out by actually performing the actions one after the other on an equilateral triangle using `GeoGebra`. The method we present works out the composite of the two symmetries namely R_{120} and F_A. Now, $R_{120} \circ F_A(A) = R_{120}(F_A(A)) = R_{120}(A) = B$. Similarly, $R_{120} \circ F_A(B) = R_{120}(C) = A$ and $R_{120} \circ F_A(C) = R_{120}(B) = C$. Hence $R_{120} \circ F_A = (A\ B)(C) = F_C$.

Row ∘ Col	R_0	R_{120}	R_{240}	F_A	F_B	F_C
R_0	R_0	R_{120}	R_{240}	F_A	F_B	F_C
R_{120}	R_{120}	R_{240}	R_0	F_C		
R_{240}	R_{240}	R_0	R_{120}			
F_A	F_A	F_B		R_0		
F_B	F_B				R_0	
F_C	F_C					R_0

Fig. B.7.1 Cayley table for the symmetry group of an equilateral triangle

(b) A non-square rectangle X has exactly four symmetries two reflections and two rotations. So the Cayley table for its symmetry group will be a 4×4 table. In this case you will find that the order in which the symmetries are composed doesn't matter. That is, if f and g are symmetries of X, then $f \circ g = g \circ f$. In a group, if the order in which the elements are combined does not matter then it is called an abelian group. So if X is a non-square rectangle then $\mathrm{Sym}(X)$ is an abelian group with 4 elements. Such an X has symmetry type D_2.

(c) Imagine an equilateral triangle whose reflection symmetries are spoilt by adding a small line segment to its vertices in such a way that the rotational symmetries are preserved. Then such an object X would have a symmetry group with exactly 3 rotational symmetries and no reflections. See Fig. B.7.2 for an illustration. This object X is said to have symmetry type C_3 as its symmetry group is a cyclic group with 3 elements. A group G is called *cyclic* if there exists an element $a \in G$ such that $G = \{a^n \mid n \in \mathbb{Z}\}$ where $a^2 = a * a, a^3 = a^2 * a, \ldots, a^0 = e$ the identity of G and $a^{-n} = (a^n)^{-1}$ for each natural number n and where g^{-1} denotes the inverse of the element g.

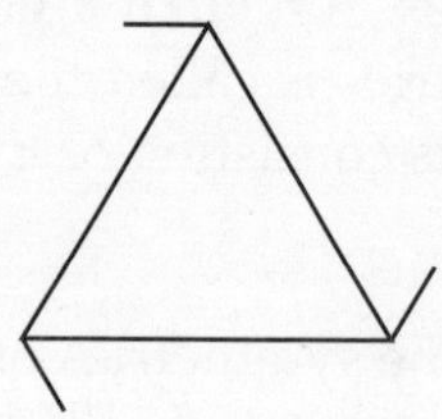

Fig. B.7.2 Finite planar figure with only rotational symmetries

In exactly the same way as above, line segments can be added to a regular n-gon in such a way that rotational symmetries are preserved but no reflection symmetries remain. Such an object would be of symmetry type C_n. An interesting result is that any finite symmetry group of a finite planar figure is of symmetry type D_n or C_n for some natural number n.

A circle is a finite planar figure with infinitely many rotational and reflection symmetries. Any angle of rotation about the centre gives a rotational symmetry and any diameter works as a line of reflection to give a reflection symmetry.

Let X be the object given in Fig. B.7.2. Then $\mathrm{Sym}(X) = \{R_0, R_{120}, R_{240}\}$. The top left corner of the Cayley table in Fig. B.7.1 is the required Cayley table.

Investigation 7.3

(a) Let R be a rotation by an angle α about the point p in the anti-clockwise direction and let R' be a rotation by an angle β about the point p' in the anti-clockwise direction. Then $R \circ R'$ is a rotation except when $\alpha + \beta$ is a multiple of 360 degree in which case it is a translation.

To find the new rotocenter in the former case, choose two points, say A and B, on the image on which the rotations R and R' are applied in succession (R' being first). Once both rotations are applied, identify the corresponding points A' and B' on the final position of the image. Take the line segments AA' and BB'. Let l and m be their perpendicular bisectors respectively. Then the point of intersection of l and m will be the new rotocenter. Can you use co-ordinate geometry or geometry to prove that this will indeed give you the new rotocenter?

(b) Let F and F' be reflections about lines l and l' respectively. Then $F \circ F'$ will be a translation if l and l' are parallel. If l and l' intersect then we get a rotation. Can you find the direction of the translation in the former case and the rotocenter and angle of rotation in the latter case?

(c) Let R be a rotation about a point p and F be a reflection about line l. If p lies on l then $R \circ F$ will be a reflection. If p does not lie on l then we get a glide reflection. Can you find the lines of reflection and glide reflection?

(d) If T is a translation about the vector v and R is a rotation about the point p then $T \circ R$ will again be a rotation.

Investigation 7.4 Note that any symmetry of Fig. 7.3.1 is nothing but a symmetry of the plane taking the strip pattern back to itself. So, we can use the knowledge from the previous investigation here.

(a) Let R and R' be rotations about rotocenters p and p' respectively. We know that for R and R' to be symmetries of the given strip pattern, the angle of rotation has to be 180 degree. So the sum of the two angles of rotation will be 360 degree and so $R \circ R'$ will be a translation.

(b) Let F and F' be reflections about lines l and l' respectively. If both the lines of reflection are vertical then they are parallel to each other and so $F \circ F'$ will be a translation. If l is a vertical line and l' is the horizontal line then $F \circ F'$ will be a rotation as the two lines of reflection intersect.

(c) If R is a rotation about a point p and F is a reflection about line l. If p lies on l then $R \circ F$ will be a reflection. So if l is horizontal then we will certainly have a reflection as all rotocenters lie on it. If l is vertical and p does not lie on l then $R \circ F$ will be a glide reflection.

(d) Let T be a translation about the vector v. Then note that v is in the direction which is horizontal. Let R be a rotation about the point p and let F be a reflection about a vertical line l. Then what can you say about $T \circ R$ will be a translation. Since the line of reflection is perpendicular to the direction of translation, therefore $T \circ F$ will be a reflection. This is the case for symmetries of a plane too.

Investigation 7.5

(b) Figure 7.4.10 has vertical lines of symmetry. It has no horizontal line of reflection symmetry. It has no 180 degree rotation and so it is of type SP Pattern III.

Figure 7.4.11 has vertical lines of symmetry, a horizontal line of reflection symmetry. So it is of type SP Pattern VII.

Investigation 7.6

(a) No.

(b) 180 and 270 degree. No. No.

(c) Yes, along the vertical direction.

Investigation 7.7 All wallpaper patterns have translations in two independent directions. Starting from WPSQ, in a clockwise direction, let us call the wallpaper patterns in Fig. 7.5.3 as WPHX, WPLL, WPRT and WPRH respectively.

WPHX has the smallest non-zero degree rotation of 120 degree. It has reflections and glide reflections. WPLL only has translations and the zero degree rotational symmetries. WPRT has the smallest non-zero degree rotation of 180 degree, reflections and glide reflections. WPRH has no non-zero rotations. But it has reflections and glide reflections.

B.8 CHAPTER 8

Investigation 8.2 The dominance matrix corresponding to the given dominance directed graph is

$$M = \begin{bmatrix} 0 & 1 & 1 & 0 & 1 \\ 0 & 0 & 1 & 1 & 0 \\ 0 & 0 & 0 & 1 & 1 \\ 1 & 0 & 0 & 0 & 1 \\ 0 & 1 & 0 & 0 & 0 \end{bmatrix}$$

Investigation 8.3 The directed graph corresponding to the given matrix is

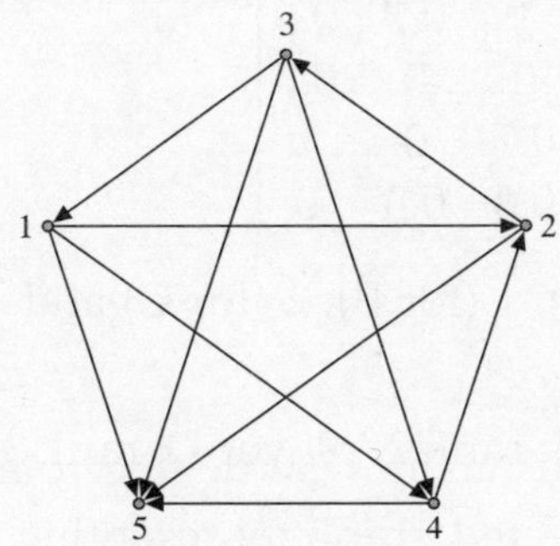

Fig. B.8.1 A directed graph

Since each pair of vertices is connected by a unique directed edge, this is a dominance directed graph.

Investigation 8.4 Each second-order win is described by a 2-edge path.

Second-order wins	2-edge paths
A over B	A-C-B
A over D	A-C-D, A-B-D
B over A	B-D-A
C over A	C-D-A
C over D	C-B-D
D over B	D-A-B
D over C	D-A-C

Investigation 8.5 The dominance matrix and its powers are:

$$M = \begin{bmatrix} 0 & 1 & 1 & 0 \\ 0 & 0 & 0 & 1 \\ 0 & 1 & 0 & 1 \\ 1 & 0 & 0 & 0 \end{bmatrix}, \quad M^2 = \begin{bmatrix} 0 & 1 & 0 & 2 \\ 1 & 0 & 0 & 0 \\ 1 & 0 & 0 & 1 \\ 0 & 1 & 1 & 0 \end{bmatrix}, \quad M^3 = \begin{bmatrix} 2 & 0 & 0 & 1 \\ 0 & 1 & 1 & 0 \\ 1 & 1 & 1 & 0 \\ 0 & 1 & 0 & 2 \end{bmatrix}.$$

A non-zero entry of M^3 is formed by combining non-zero entries of M and M^2, that is, by following an edge with a 2-edge. Therefore, $m_{ij}^{(3)}$ represents the number of 3-edge paths from player i to player j.

For example, $m_{23}^{(3)} = 1$ indicates a single 3-edge path from B to C. This 3-edge path is B-D-A-C.

Investigation 8.6

$$0.7M + 0.3M^2 = \begin{bmatrix} 0 & 1 & 0.7 & 0.6 \\ 0.3 & 0 & 0 & 0.7 \\ 0.3 & 0.7 & 0 & 1 \\ 0.7 & 0.3 & 0.3 & 0 \end{bmatrix}$$

The row sums are 2.3 (for A), 1 (for B), 2 (for C) and 1.3 (for D). Accordingly, the ranks are

Player A: rank 1, player C: rank 2, player D: rank 3 and player B: rank 4.

Investigation 8.7 (a) The directed graph representing the flight connections is

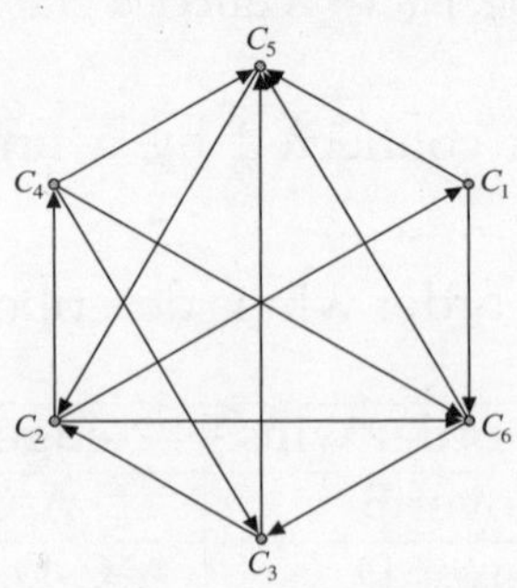

Fig. B.8.2 A directed graph showing flight connections

(b) The dominance matrix and its powers are

$$M = \begin{bmatrix} 0 & 0 & 0 & 0 & 1 & 1 \\ 1 & 0 & 0 & 1 & 0 & 1 \\ 0 & 1 & 0 & 0 & 1 & 0 \\ 0 & 0 & 1 & 0 & 1 & 1 \\ 0 & 1 & 0 & 0 & 0 & 0 \\ 0 & 0 & 1 & 0 & 1 & 0 \end{bmatrix}, M^2 = \begin{bmatrix} 0 & 1 & 1 & 0 & 1 & 0 \\ 0 & 0 & 2 & 0 & 3 & 2 \\ 1 & 1 & 0 & 1 & 0 & 1 \\ 0 & 2 & 1 & 0 & 2 & 0 \\ 1 & 0 & 0 & 1 & 0 & 1 \\ 0 & 2 & 0 & 0 & 1 & 0 \end{bmatrix}, M^3 = \begin{bmatrix} 0 & 1 & 1 & 0 & 2 & 1 \\ 1 & 0 & 2 & 1 & 3 & 3 \\ 1 & 2 & 0 & 1 & 1 & 1 \\ 0 & 2 & 2 & 0 & 3 & 1 \\ 1 & 1 & 0 & 1 & 0 & 1 \\ 0 & 2 & 1 & 0 & 2 & 0 \end{bmatrix}$$

(c) Only the last combination is able to distinguish among all the cities.

Investigation 8.9 The probabilities of the offspring genotypes are as follows.

	AA			Aa			aa		
	AA	Aa	aa	AA	Aa	aa	AA	Aa	aa
AA	1	1/2	0	1/2	1/4	0	0	0	0
Aa	0	1/2	1	1/2	1/2	1/2	1	1/2	0
aa	0	0	0	0	1/4	1/2	0	1/2	1

Investigation 8.10 The fundamental genetic matrices, which refer to mating with genotypes **Aa** and **aa** are $\begin{bmatrix} 1/2 & 1/4 & 0 \\ 1/2 & 1/2 & 1/2 \\ 0 & 1/4 & 1/2 \end{bmatrix}$ and $\begin{bmatrix} 0 & 0 & 0 \\ 1 & 1/2 & 0 \\ 0 & 1/2 & 1 \end{bmatrix}$ respectively.

Investigation 8.11

(a) The distribution of the first generation offspring when all plants are fertilised with **Aa**.

$$\begin{bmatrix} 1/2 & 1/4 & 0 \\ 1/2 & 1/2 & 1/2 \\ 0 & 1/4 & 1/2 \end{bmatrix} \begin{bmatrix} p \\ q \\ r \end{bmatrix} = \begin{bmatrix} (2p+q)/4 \\ 1/2 \\ (2r+q)/4 \end{bmatrix}$$

(b) The distribution of the first generation offspring when all plants are fertilised with **aa**.

$$\begin{bmatrix} 0 & 0 & 0 \\ 1 & 1/2 & 0 \\ 0 & 1/2 & 1 \end{bmatrix} \begin{bmatrix} p \\ q \\ r \end{bmatrix} = \begin{bmatrix} 0 \\ p+q/2 \\ r+q/2 \end{bmatrix}$$

Investigation 8.12 Apply Mathematical Induction.

Investigation 8.13 We have

$$M = \begin{bmatrix} 1/2 & 1/4 & 0 \\ 1/2 & 1/2 & 1/2 \\ 0 & 1/4 & 1/2 \end{bmatrix} \quad \text{and} \quad x^{(0)} = \begin{bmatrix} 0.1 \\ 0.2 \\ 0.7 \end{bmatrix}$$

The `Excel` computations given below show that in the 5th generation, about 25% of the plant population will have genotype **AA**, 50% will have genotype **Aa**, while 25% will have genotype **aa**. Thus, approximately 75% of the plant population will have red flowers and the remaining 25% will have white flowers.

B	C	D	E	F	G	H	I	J	K
	M			x(0)	x(1)	x(2)	x(3)	x(4)	x(5)
0.5	0.25	0		0.1	0.1	0.175	0.2125	0.23125	0.240625
0.5	0.5	0.5		0.2	0.5	0.5	0.5	0.5	0.5
0	0.25	0.5		0.7	0.4	0.325	0.2875	0.26875	0.259375

Fig. B.8.3 Genotype distributions up to five generations

Investigation 8.14 Plots of the changes in genotype distributions:

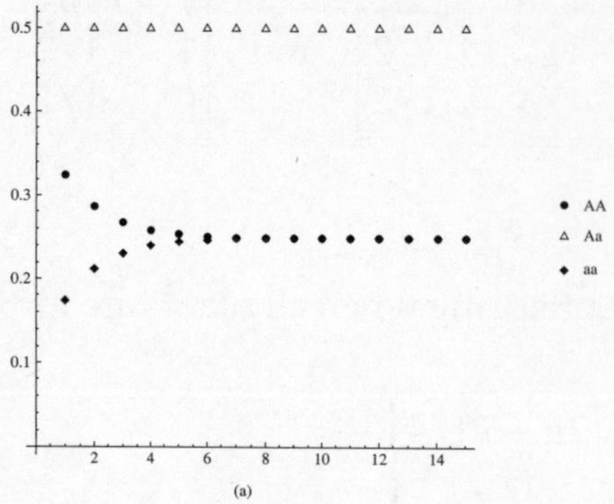

(a)

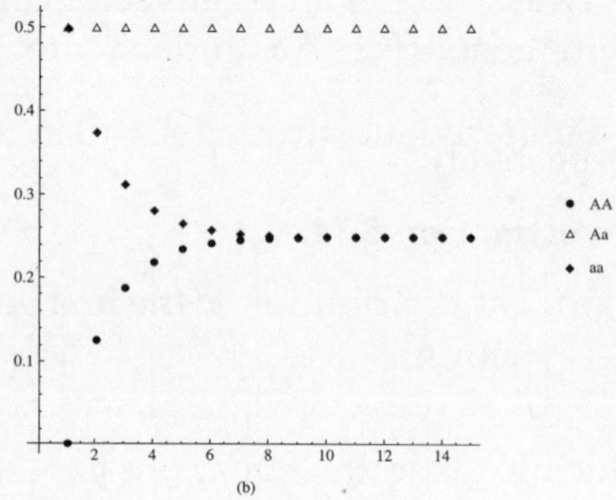

(b)

Fig. B.8.4 Plots of genotype distributions

In both cases, the final proportions are 50% of **Aa** and 25% each of **AA** and **aa**. Hence the final phenotype distributions are 75% red flowers and 25% white flowers. Do you see anything curious about the **Aa** trend?

Investigation 8.15 Apply Mathematical Induction.

Investigation 8.16 $\begin{vmatrix} 1-\lambda & \frac{1}{2} & 0 \\ 0 & \frac{1}{2}-\lambda & 1 \\ 0 & 0 & -\lambda \end{vmatrix} = 0$ simplifies to $(1-\lambda)(\frac{1}{2}-\lambda)(-\lambda) = 0$.

This leads to $\lambda = 0, 1/2, 1$ as the eigenvalues of the matrix M.

Investigation 8.17 Substitute $\lambda = 1/2$ in the matrix equation $(M - \lambda I)X = O$:

$$\begin{bmatrix} 1-\lambda & 1/2 & 0 \\ 0 & 1/2-\lambda & 1 \\ 0 & 0 & -\lambda \end{bmatrix} \begin{bmatrix} x_1 \\ x_2 \\ x_3 \end{bmatrix} = \begin{bmatrix} 0 \\ 0 \\ 0 \end{bmatrix}$$

We get

$$\begin{bmatrix} 1/2 & 1/2 & 0 \\ 0 & 0 & 1 \\ 0 & 0 & -1/2 \end{bmatrix} \begin{bmatrix} x_1 \\ x_2 \\ x_3 \end{bmatrix} = \begin{bmatrix} 0 \\ 0 \\ 0 \end{bmatrix}$$

which leads to the equations $x_1 + x_2 = 0$ and $x_3 = 0$. We can make an arbitrary choice for x_1. Let $x_1 = -1$. Then $x_2 = 1$ and $x_3 = 0$.

Thus $\begin{bmatrix} -1 \\ 1 \\ 0 \end{bmatrix}$ is an eigenvector corresponding to the eigenvalue $\lambda = 1/2$.

Similarly, substituting $\lambda = 0$ in the matrix equation $(M - \lambda I)X = O$ we get

$$\begin{bmatrix} 1 & 1/2 & 0 \\ 0 & 1/2 & 1 \\ 0 & 0 & 0 \end{bmatrix} \begin{bmatrix} x_1 \\ x_2 \\ x_3 \end{bmatrix} = \begin{bmatrix} 0 \\ 0 \\ 0 \end{bmatrix}$$

which leads to the equations $x_1 + \frac{1}{2}x_2 = 0$ and $\frac{1}{2}x_2 + x_3 = 0$. Let $x_3 = 1$. Then $x_2 = -2$ and $x_1 = 1$.

Thus $\begin{bmatrix} 1 \\ -2 \\ 1 \end{bmatrix}$ is an eigenvector corresponding to the eigenvalue $\lambda = 0$.

Investigation 8.18 We have

$$x^{(0)} = \alpha \begin{bmatrix} 1 \\ 0 \\ 0 \end{bmatrix} + \beta \begin{bmatrix} -1 \\ 1 \\ 0 \end{bmatrix} + \gamma \begin{bmatrix} 1 \\ -2 \\ 1 \end{bmatrix} . = \begin{bmatrix} \alpha - \beta + \gamma \\ \beta - 2\gamma \\ \gamma \end{bmatrix}$$

The requirement that the entries of $x^{(0)}$ have to sum to one gives $\alpha = 1$. For (b),

$$M^n x^{(0)} = 1^n \begin{bmatrix} 1 \\ 0 \\ 0 \end{bmatrix} + \left(\frac{1}{2}\right)^n \beta \begin{bmatrix} -1 \\ 1 \\ 0 \end{bmatrix} + 0^n \gamma \begin{bmatrix} 1 \\ -2 \\ 1 \end{bmatrix} \to \begin{bmatrix} 1 \\ 0 \\ 0 \end{bmatrix}.$$

Investigation 8.19 Fertilise the red flowered plants with the white flowered ones. Suppose the initial fractions of **AA** and **Aa** among the red flowered plants were α and $1 - \alpha$ then, the probability vector for the next generation will be

$$\begin{bmatrix} 0 & 0 & 0 \\ 1 & 1/2 & 0 \\ 0 & 1/2 & 1 \end{bmatrix} \begin{bmatrix} \alpha \\ 1 - \alpha \\ 0 \end{bmatrix} = \begin{bmatrix} 0 \\ (1 + \alpha)/2 \\ (1 - \alpha)/2 \end{bmatrix}$$

If the red flowers make a fraction β of the first generation, we get $\alpha = 2\beta - 1$.

Investigation 8.20 The steady state distribution for the case when all plants are fertilised with plants of genotype **Aa** is $\begin{bmatrix} 1/4 \\ 1/2 \\ 1/2 \end{bmatrix}$. Thus, in the long run 75% of the plants will have red flowers and the remaining 25% will have white flowers. This will happen irrespective of the initial distribution.

Investigation 8.22 This process is governed by the matrix $M = \begin{bmatrix} 1 & 1/4 & 0 \\ 0 & 1/2 & 0 \\ 0 & 1/4 & 1 \end{bmatrix}$.

The compositions after 20 generations for each initial population are as follows:

$$\text{(a)} \begin{bmatrix} 0.1 \\ 0.8 \\ 0.1 \end{bmatrix} \rightarrow \begin{bmatrix} 0.5 \\ 0 \\ 0.5 \end{bmatrix} \qquad \text{(b)} \begin{bmatrix} 0.8 \\ 0.1 \\ 0.1 \end{bmatrix} \rightarrow \begin{bmatrix} 0.85 \\ 0 \\ 0.15 \end{bmatrix} \qquad \text{(c)} \begin{bmatrix} 0.4 \\ 0.4 \\ 0.2 \end{bmatrix} \rightarrow \begin{bmatrix} 0.6 \\ 0 \\ 0.4 \end{bmatrix}$$

Investigation 8.23 The eigenvalues of M in the previous Investigation are $1, 1, 0.5$, with corresponding eigenvectors $\begin{bmatrix} 1 \\ 0 \\ 0 \end{bmatrix}$, $\begin{bmatrix} 0 \\ 0 \\ 1 \end{bmatrix}$ and $\begin{bmatrix} -1 \\ 2 \\ -1 \end{bmatrix}$. Any initial $x^{(0)} = \begin{bmatrix} p \\ q \\ r \end{bmatrix}$ can be expressed in terms of the eigenvectors as follows:

$$\begin{bmatrix} p \\ q \\ r \end{bmatrix} = (p + \frac{q}{2}) \begin{bmatrix} 1 \\ 0 \\ 0 \end{bmatrix} + (r + \frac{q}{2}) \begin{bmatrix} 0 \\ 0 \\ 1 \end{bmatrix} + \frac{q}{2} \begin{bmatrix} -1 \\ 2 \\ -1 \end{bmatrix}$$

Hence, as $n \to \infty$,

$$M^n x^{(0)} \rightarrow (p + \frac{q}{2}) \begin{bmatrix} 1 \\ 0 \\ 0 \end{bmatrix} + (r + \frac{q}{2}) \begin{bmatrix} 0 \\ 0 \\ 1 \end{bmatrix} = \begin{bmatrix} p + q/2 \\ 0 \\ r + q/2 \end{bmatrix}$$

Investigation 8.25

		Parent pairings					
		(A,AA)	**(A,Aa)**	**(A,aa)**	**(a,AA)**	**(a,Aa)**	**(a,aa)**
Male offspring	**A**	1	1/2	0	1	1/2	0
	a	0	1/2	1	0	1/2	1
Female offspring	**AA**	1	1/2	0	0	0	0
	Aa	0	1/2	1	1	1/2	0
	aa	0	0	0	0	1/2	1

Investigation 8.27 Using the equations

$$X^{(1)} = LX^{(0)},\ X^{(2)} = LX^{(1)},\ X^{(3)} = LX^{(2)},\ X^{(4)} = LX^{(3)},\ X^{(5)} = LX^{(4)}$$

and computing using `Excel` we get

				X(0)	X(1)	X(2)	X(3)	X(4)	X(5)
1	4	3	2	100	1000	1349	3415	6616.3	14196.5
0.5	0	0	0	100	50	500	674.5	1707.5	3308.13
0	0.33	0	0	100	33	16.5	165	222.59	563.475
0	0	0.25	0	100	25	8.25	4.125	41.25	55.6463

Fig. B.8.5 Excel computation representing the population in the four age groups

After 30 years, the population in the four age groups will be 3992, 586, 117 and 15 respectively.

Investigation 8.29

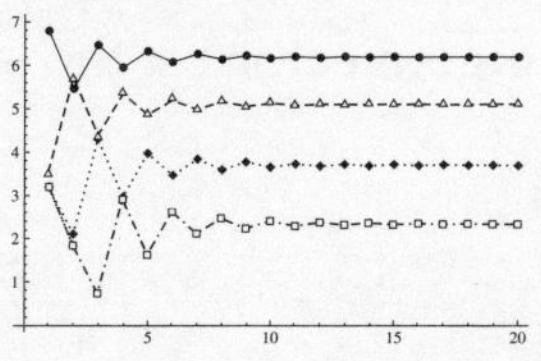

(a)

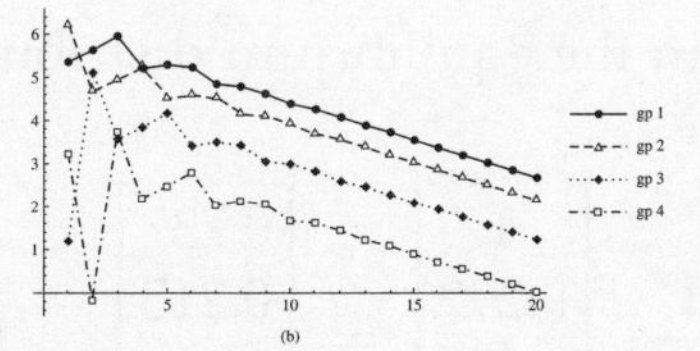

(b)

Fig. B.8.6 Plots representing the population growth in the four age groups

The first population becomes constant over time, the second becomes extinct. If you try different starting distributions, you will see that the long term trends do not depend on the initial state.

Investigation 8.30 We present the proof for the 3×3 case. The reader should extend it to the general $n \times n$ case. First, we expand the $Lv_1 = \lambda_1 v_1$ equation:

$$\begin{bmatrix} k_1 & k_2 & k_3 \\ p_1 & 0 & 0 \\ 0 & p_2 & 0 \end{bmatrix} \begin{bmatrix} x_1 \\ x_2 \\ x_3 \end{bmatrix} = \lambda_1 \begin{bmatrix} x_1 \\ x_2 \\ x_3 \end{bmatrix}$$

Equating the corresponding elements on both sides, we get

$$\begin{aligned} k_1x_1 + k_2x_2 + k_3x_3 &= \lambda_1 x_1 \\ p_1x_1 &= \lambda_1 x_2 \\ p_2x_2 &= \lambda_1 x_3 \end{aligned}$$

We note that if $x_1 = 0$ then $x_2 = x_3 = 0$ and hence $v_1 = 0$. So we must have $x \neq 0$. We set $x_1 = 1$ This gives $x_2 = p_1/\lambda_1$ and $x_3 = p_2p_1/\lambda_1^2$. Thus,

$$v_1 = \begin{bmatrix} 1 \\ p_1/\lambda_1 \\ p_2p_1/\lambda_1^2 \end{bmatrix}$$

Investigation 8.31

We have $L = \begin{bmatrix} 1 & 2 & 2 & 2 \\ 1/2 & 0 & 0 & 0 \\ 0 & 1/3 & 0 & 0 \\ 0 & 0 & 1/4 & 0 \end{bmatrix}$. The dominant eigenvalue is $\lambda_1 = 1.714$.

A corresponding eigenvector is $v_1 = \begin{bmatrix} 1 \\ p_1/\lambda_1 \\ p_1 p_2/\lambda_1^2 \\ p_1 p_2 p_3/\lambda_1^3 \end{bmatrix} = \begin{bmatrix} 1 \\ 0.292 \\ 0.057 \\ 0.008 \end{bmatrix}$.

To obtain the equilibrium distribution, we divide this eigenvector by the sum of its entries:

$$\frac{1}{1.357} \begin{bmatrix} 1 \\ 0.292 \\ 0.057 \\ 0.008 \end{bmatrix} = \begin{bmatrix} 0.737 \\ 0.215 \\ 0.042 \\ 0.006 \end{bmatrix}$$

Bibliography

[1] *Alhambra of Granada*. URL: https://www.alhambradegranada.org/en/info/galleryofphotographs/tiles.asp.

[2] Howard Anton and Chris Rorres. *Elementary Linear Algebra*. 11th ed. International Student Edition. Wiley, 2016.

[3] Shobha Bagai, Amber Habib and Geetha Venkataraman. *A Bridge to Mathematics*. Sage Publications, India, 2017.

[4] N L Biggs, E K Lloyd and R J Wilson. *Graph Theory 1736–1936*. Clarendon Press Oxford, 1998.

[5] A Bunde and S Havlin. A brief introduction to fractal geometry. *Fractals in Science*. 3rd ed. Springer-Verlag, 1994, pp. 18–19.

[6] David Burton. *Elementary Number Theory*. 7th ed. McGraw Hill Education, 2017.

[7] Davide Castelvecchi. Quantum-computing pioneer warns of complacency over Internet security. *Nature*, 587 (Nov. 2020), p. 189.

[8] R Derousseau. The best bets at the casino. *Forbes Magazine*, 2007.

[9] Whitfield Diffe and Martin E Hellman. New Directions in Cryptography. *IEEE Transactions on Information Theory*, IT-22.6 (Nov. 1976).

[10] Marcus du Sautoy. *Symmetry: A Journey into the Patterns of Nature*. Harper Collins, 2008.

[11] David Eck. *Mandelbrot Viewer*. URL: https://math.hws.edu/eck/js/mandelbrot/MB.html.

[12] Leonhard Euler. Solutio problematis ad geometriam situs pertinentis. *Commentarii Academiae Scientiarum Imperialis Petropolitanae, Comm. Acad. Sci. Imp. Petropol* 8 (1736). = Opera Omnia (I) 7 (191–156), 1–10, pp. 128–140. URL: http://eulerarchive.maa.org//docs/originals/E053.pdf.

[13] *Exploring symmetries of Platonic Solids using GeoGebra*. URL: www.geogebra.org/m/eJdY6ckN.

[14] Ruma Falk. *Understanding Probability and Statistics: A book of problems*. Universities Press (India), 1992.

[15] Joseph A Gallian. Contemporary Abstract Algebra. 8th ed. Brookes/Cole, Cengage Learning, 2013.

[16] *GeoGebra Tutorials*. URL: https://www.geogebra.org/a/14.

[17] J Ghosh. Exploring fractals—The GeoGebra way. *At Right Angles: A resource for school mathematics*, 4 (2019), pp. 77–83.

[18] Richard Gill. The Monty Hall Problem is not a probability puzzle (it's a challenge in mathematical modelling). *Statistica Neerlandica*, 65.1 (2011), pp. 58–71.

[19] Brian Hopkins and Robin J Wilson. The Truth about Königsberg. *The College Mathematics Journal*, 35.3 (May 2004), pp. 198–207.

[20] B Houston and H Masum. *Explorations in 4-peg Tower of Hanoi*. 2004. URL: http://service.scs.carleton.ca/sites/default/files/tr/TR-04-10.pdf.

[21] Amitabh Joshi. Hardy–Weinberg Equilibrium and the Foundations of Evolutionary Genetics, Part 1: The Dance of the Genes. *Resonance*, 13.9 (2008), pp. 812– 835.

[22] Amitabh Joshi. Hardy–Weinberg Equilibrium and the Foundations of Evolutionary Genetics, Part 2: The Inertial State in Population Genetics. *Resonance*, 13.10 (2008), pp. 951–970.

[23] Amitabh Joshi. Hardy–Weinberg Equilibrium and the Foundations of Evolutionary Genetics, Part 3: Incorporating Mutation and Migration. *Resonance*, 16.2 (2011), pp. 116–128.

[24] J N Kapur. *Mathematical Models in Biology and Medicine*. Affiliated East-West Press, 1985.

[25] David Kuncicky and Ronald Larsen. Introduction to Excel. 5th ed. Pearson Prentice Hall, 2013.

[26] L P Lefkovitch. The Study of Population Growth in Organisms Grouped by Stages. *Biometrics*, 21.1 (1965), pp. 1–18.

[27] P H Leslie. On the Use of Matrices in Certain Population Mathematics. *Biometrika*. 33.3 (1945), pp. 183–212.

[28] Steven Levy. The Open Secret. *Wired* (1999). URL: https://www.wired.com/1999/04/crypto/.

[29] Roger C Lyndon. *Groups and Geometry*. London Mathematical Society Lecture Note Series. Cambridge University Press, 1985.

[30] Benoit B Mandelbrot. *Fractals: Form, chance, and dimension*. W H Freeman and Co., 1977.

[31] Benoit B Mandelbrot. *The Fractal Geometry of Nature*. W H Freeman and Co., 1982.

[32] The Mathemagicians' Guild. *The Mandelbrot Set Explained Series*. 2020. URL: https://youtube.com/playlist?list=PL9tHLTl03LqG4ajDvqyfCDMKSxmR_plJ3.

[33] Mark McClure. Newton's method for complex polynomials. *Mathematica in Education and Research*, 11.2 (2006).

[34] Math Warehouse. *Monty Hall Simulation Online*. URL: https://www.mathwarehouse.com/monty-hall-simulation-online.

[35] *M C Escher Foundation*. URL: https://mcescher.com.

[36] Dan Ma. *The gambler's ruin problem*. URL: https://probabilityandstats.wordpress.com/category/game-of-chance/.

[37] J Metz. Twists on the Tower of Hanoi. *The Mathematics Teacher*, 107.9 (2014), pp. 712–715.

[38] Barry Nalebuff. Puzzles: Choose a curtain, duel-ity, two point conversions, and more. *J. Econ. Perspectives*, 1.2 (1987), pp. 157–163.

[39] J R Newman, ed. *The World of Mathematics*, Vol. 1. Simon and Schuster, 1956.

[40] J J O'Connor and E F Robertson. *Jacob (Jacques) Bernoulli*. 1998. URL: https://mathshistory.st-andrews.ac.uk/Biographies/Bernoulli_Jacob/.

[41] J J O'Connor and E F Robertson. *Georges Louis Leclerc Comte de Buffon*. 2004. URL: https://mathshistory.st-andrews.ac.uk/Biographies/Buffon/.

[42] J J O'Connor and E F Robertson. *Arthur Cayley*. 2014. URL: https://mathshistory.st-andrews.ac.uk/Biographies/Cayley/.

[43] J J O'Connor and E F Robertson. *Pafnuty Lvovich Chebyshev*. 2002. URL: https://mathshistory.st-andrews.ac.uk/Biographies/Chebyshev/.

[44] J J O'Connor and E F Robertson. *Maurits Cornelius Escher*. 2000. URL: https://mathshistory.st-andrews.ac.uk/Biographies/Escher/.

[45] J J O'Connor and E F Robertson. *Euclid of Alexandria*. 1999. URL: https://mathshistory.st-andrews.ac.uk/Biographies/Euclid/.

[46] J J O'Connor and E F Robertson. *Leonhard Euler*. 1998. URL: https://mathshistory.st-andrews.ac.uk/Biographies/Euler/.

[47] J J O'Connor and E F Robertson. *Pierre Joseph Louis Fatou*. 2016. URL: https://mathshistory.st-andrews.ac.uk/Biographies/Fatou/.

[48] J J O'Connor and E F Robertson. *Pierre de Fermat*. 1996. URL: https://mathshistory.st-andrews.ac.uk/Biographies/Fermat/.

[49] J J O'Connor and E F Robertson. *Sir William Rowan Hamilton*. 1998. URL: https://mathshistory.st-andrews.ac.uk/Biographies/Hamilton/.

[50] J J O'Connor and E F Robertson. *Gaston Maurice Julia*. 2008. URL: https://mathshistory.st-andrews.ac.uk/Biographies/Julia/.

[51] J J O'Connor and E F Robertson. *Niels Fabian Helge von Koch*. 2000. URL: https://mathshistory.st-andrews.ac.uk/Biographies/Koch/.

[52] J J O'Connor and E F Robertson. *Joseph-Louis Lagrange*. 1999. URL: https://mathshistory.st-andrews.ac.uk/Biographies/Lagrange/.

[53] J J O'Connor and E F Robertson. *Edouard Lucas*. 1996. URL: https://mathshistory.st-andrews.ac.uk/Biographies/Lucas/.

[54] J J O'Connor and E F Robertson. *Benoit Mandelbrot*. 1999. URL: https://mathshistory.st-andrews.ac.uk/Biographies/Mandelbrot/.

[55] J J O'Connor and E F Robertson. *Curtis Tracy McMullen*. 2009. URL: https://mathshistory.st-andrews.ac.uk/Biographies/McMullen/.

[56] J J O'Connor and E F Robertson. *Waclaw Sierpinski*. 1997. URL: https://mathshistory.st-andrews.ac.uk/Biographies/Sierpinski/.

[57] J J O'Connor and E F Robertson. *James Joseph Sylvester*. 2005. URL: https://mathshistory.st-andrews.ac.uk/Biographies/Sylvester/.

[58] H O Peitgen, D Saupe and F V Haeseler. Cayley's Problem and Julia Sets. *The Mathematical Intelligencer*, 6.2 (1984), pp. 11–20.

[59] J F Ramaley. Buffon's Noodle Problem. *The American Mathematical Monthly*, 76.8 (1969), pp. 916–918.

[60] R L Rivest, A Shamir and L Adleman. A method for obtaining digital signatures and public-key cryptosystems. *Communications of the ACM*, 21.2 (1978), pp. 120–126.

[61] K H Rosen. *Discrete Mathematics and its Applications*. 4th ed. Tata McGraw-Hill, 2001.

[62] Sheldon Ross. *Introduction To Probability Models*. 9th ed. Elsevier India, 2007.

[63] Sheldon Ross. *A First Course in Probability*. 9th ed. Pearson Education, India, 2019.

[64] Steve Selvin. A problem in probability (letter to the editor). *The American Statistician*, 29.1 (1975), pp. 67–71.

[65] Leonid Shifrin. *Mathematica®Programming—an advanced introduction*. URL: http://www.mathprogramming-intro.org/.

[66] Simon Singh. *The Code Book: The Science of Secrecy from Ancient Egypt to Quantum Cryptography*. Anchor, 2000.

[67] David P Smith and Nathan Keyfitz. *Mathematical Demography: selected papers*. Kenneth W Wachter and Hervé Le Bras, Eds. 2nd revised ed. Springer, 2013.

[68] Barak Sofer. *Excel Tutorial*. 2020. URL: http://www.free-training-tutorial.com/format-cells.html.

[69] Philip D Straffin Jr. Newton's Method and Fractal Patterns. *UMAP Modules: Tools for Teaching* (1991).

[70] Gilbert Strang. A Chaotic Search for i. *College Mathematics Journal*, 22.1(1991), pp. 3–12.

[71] Yaqin Sun. A Mathematical Analysis of a Game of Craps. *Undergraduate Journal*, 10 (2014), pp. 149–154. URL: http://vc.bridgew.edu/undergrad_rev/vol10/iss1/29.

[72] Peter Taylor. *What Ever Happened to Those Bridges?* Australian Mathematics Trust, University of Canberra. 2000. URL: https://web.archive.org/web/20120319074335/http://www.amt.canberra.edu.au/koenigs.html.

[73] Bruce F Torrence and Eve A Torrence. *The Student's Introduction to MATHEMATICA®*. 2nd ed. Cambridge University Press, 2009.

[74] *Towers of Hanoi Animation*. URL: http://towersofhanoi.info/Animate.aspx.

[75] Geetha Venkataraman. Of Monsters and Moonshine, a review of 'Symmetry' by Marcus Du Sautoy. *At Right Angles*, 3.1 (Mar. 2014). URL: http://teachersofindia.org/en/article/monsters-and-moonshine.

[76] Geetha Venkataraman. Through the Symmetry Lens: Part I- Planar Symmetry and frieze Patterns. *At Right Angles*, 5.1 (Mar. 2016), pp. 5–14. URL: http://teachersofindia.org/en/ebook/through-symmetry-lens-1.

[77] Geetha Venkataraman. Through the Symmetry Lens: Part II- Wallpaper Patterns and Symmetry around us. *At Right Angles*, 5.2 (July 2016), pp. 11–16. URL: http://teachersofindia.org/en/ebook/through-symmetry-lens-part-2.

[78] Marilyn vos Savant. Ask Marilyn. *Parade* (9 September 1990), p. 16.

[79] Stan Wagon. *The Buffon Noodle Problem*. Wolfram Demonstrations Project. 2011. URL: http://demonstrations.wolfram.com/TheBuffonNoodleProblem/.

[80] *Wallpaper Symmetry Generating Application*. URL: http://math.hws.edu/eck/js/symmetry/wallpaper.html.

[81] Eric W Weisstein. *Craps*. MathWorld–A Wolfram Web Resource. URL: https://mathworld.wolfram.com/Craps.html.

[82] Eric Weisstein. Koch Antisnowflake. MathWorld–A Wolfram Web Resource. URL: https://mathworld.wolfram.com/KochAntisnowflake.html.

[83] Hermann Weyl. *Symmetry*. Princeton University Press, 1952.

[84] John Whitehouse. *Newton-Raphson Patterns*. 2015. URL: http://www.eddaardvark.co.uk/python_patterns/newton_index.html.

[85] Wikipedia. *Buffon's Needle Problem*. 2021. URL: https://en.wikipedia.org/wiki/Buffon's_needle_problem.

[86] Wikipedia. *Pythagoras tree (fractal)*. 2021. URL: https://en.wikipedia.org/wiki/Pythagoras_tree_(fractal).

[87] Wikipedia. *Quantum Computing*. 2021. URL: https://en.wikipedia.org/wiki/Quantum_computing.

[88] Wikipedia. *RSA Factoring Challenge*. 2021. URL: https://en.wikipedia.org/wiki/RSA_Factoring_Challenge.

[89] R J Wilson. *Introduction to Graph Theory*. 4th ed. Addison Wesley, 1996.

[90] Lily Yau and Adi Ben-Israel. The Newton and Halley Methods for Complex Roots. *The American Mathematical Monthly*, 105 (1998), pp. 806–818.

Image Credits

[C1] Amber Habib.

[C2] User: Ushakaron /Wikimedia Commons. Public domain.

[C3] Jonaki Ghosh.

[C4] User: Luringen /Wikimedia Commons. CC BY-SA 3.0.

[C5] Jean Claude Constantin.

[C6] Geetha Venkataraman.

[C7] Jonaki Ghosh.

[C8] Jonaki Ghosh.

[C9] User: Chfer532 /Wikimedia Commons. CC BY 3.0.

[C10] Dick Culbert /Wikimedia Commons. CC BY 2.0.

Subject Index

Index of Names